How far?

Name:

(A) Measure with cubes.

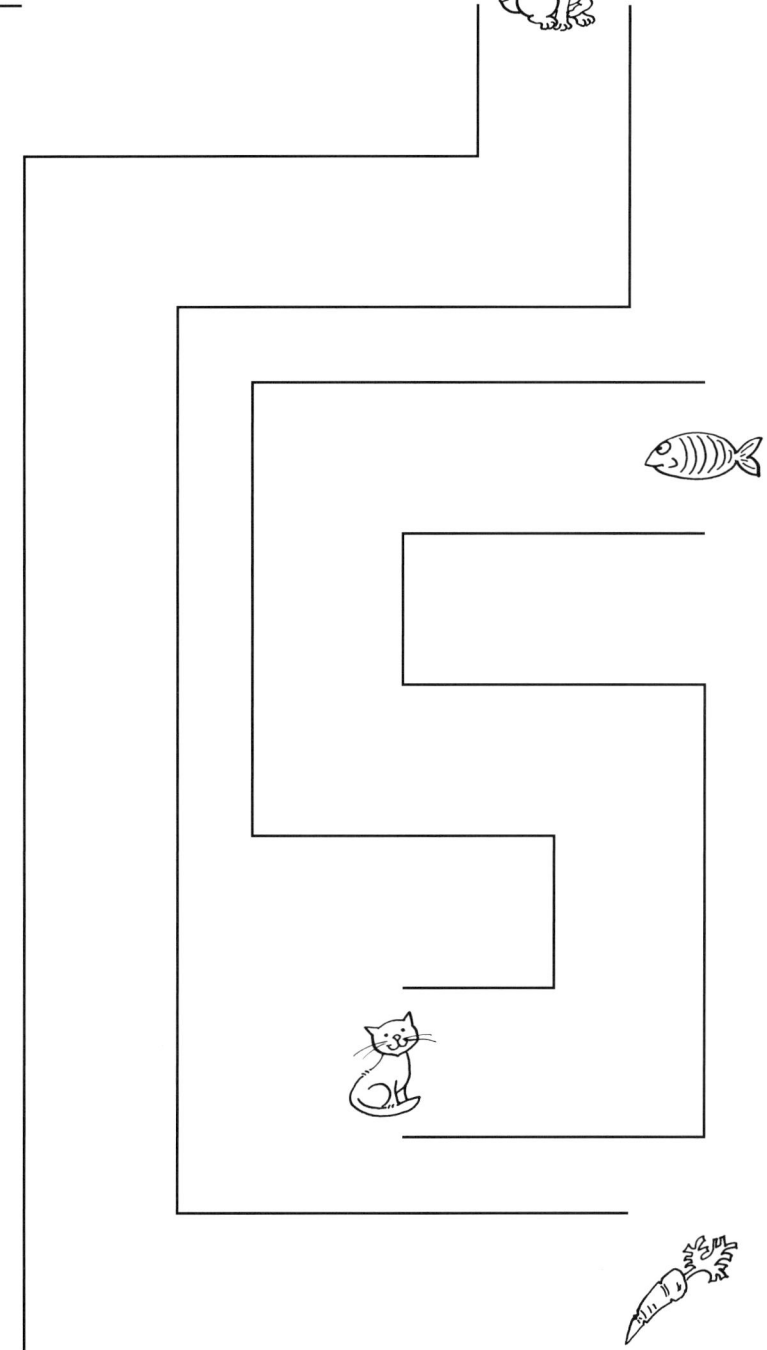

1 How far has 🐕 got to go

to get his 🦴?

☐ cubes

2 How far has 🐇 got to go to get her 🥕? ☐ cubes

3 How far has 🐱 got to go to get her 🐟? ☐ cubes

4 Who goes furthest?

Notes/date:

Bean bag throw

Name:

Furthest throw (Name and how far)	Nearest throw (Name and how far)	How much further?
Joanne 6 m and a bit	Milla 4 m and a bit	6 m − 4 m = 2 m

Names	Target		Total points
	Closest throw Score 2 points	Throw within 1 m of target Score 1 point	
Team A _____ _____ _____			
Team B _____ _____			

Notes/date:

Cambridge Mathematics Direct 2 © Cambridge University Press 2002   M1.2

Straw cottage

Name:

C   Measure the length of each dotted line in centimetres.
Cut a straw to fit, and glue it onto your picture.
Write the length of the line, in centimetres, in the box.

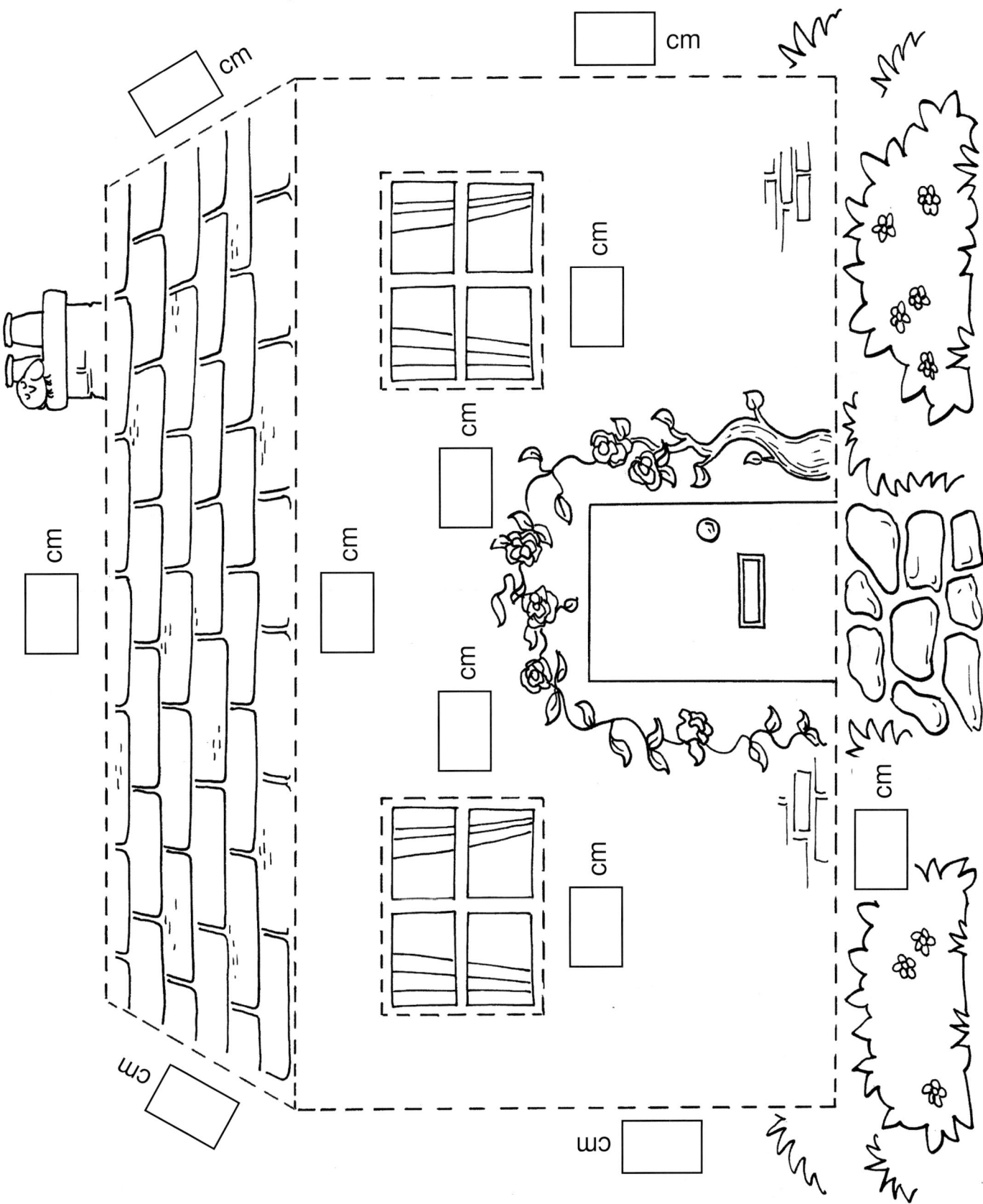

Measuring in centimetres

Name:

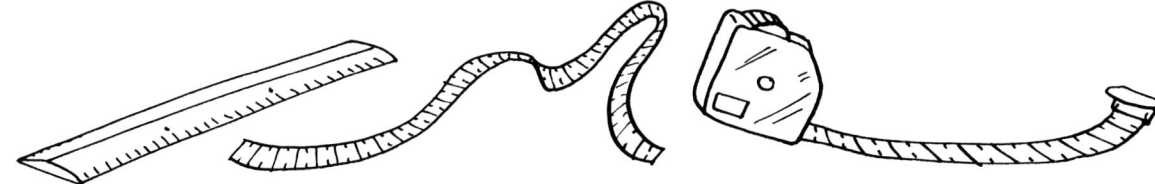

About I cm	About I0 cm	About I00 cm

Dot to dot

(B)   Use your ruler to join the dots in alphabetical order, from a to m.

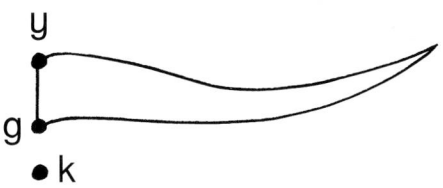

y

g

• k

• l

• •j • i
m

h •

• •f
a

x
•

• e

•
b

•
c

•
d

1   How long is the boat (from b to e)?   [ ] cm

2   How deep is the boat (from d to x)?   [ ] cm

3   How wide is the big sail (from i to h)?   [ ] cm

4   How wide is the small sail (from l to j)?   [ ] cm

5   How tall is the mast (from f to y)?   [ ] cm

My body measurements

Name:

★/A My handspan is ☐ centimetres.

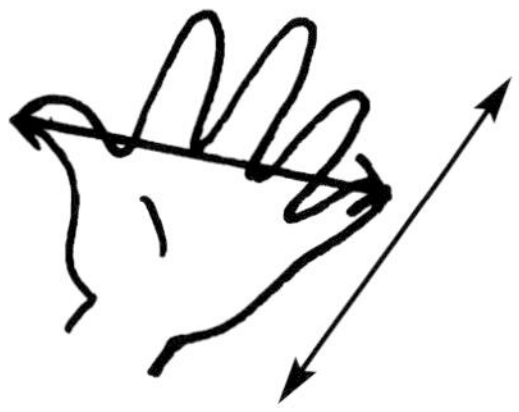

My hand is ☐ cm long.

Around my head:

estimate ☐

measure ☐

Length of my arm:

estimate ☐

measure ☐

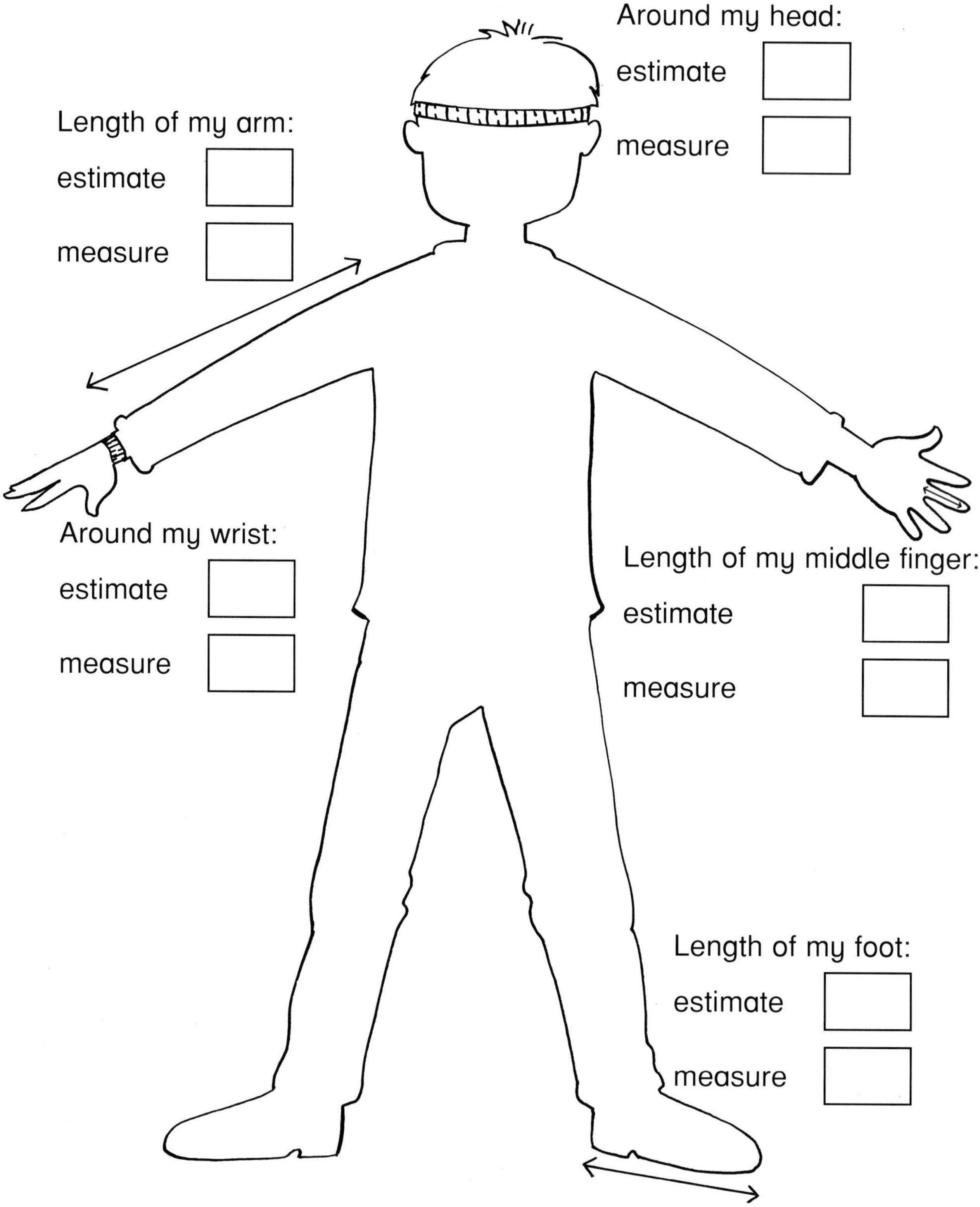

Around my wrist:

estimate ☐

measure ☐

Length of my middle finger:

estimate ☐

measure ☐

Length of my foot:

estimate ☐

measure ☐

Notes/date:

How to measure

Name:

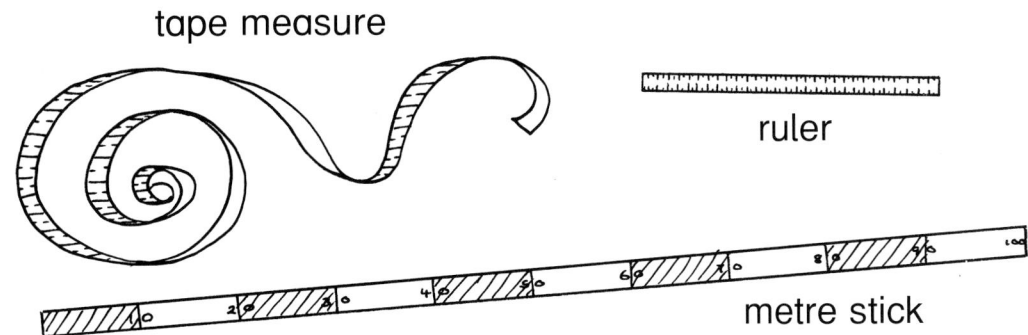

tape measure

ruler

metre stick

(B) Choose what you would use, then measure each length.
Measure in metres or centimetres. Estimate first.

What I measured	Ruler, metre stick or tape measure	Estimate	Measurement
Width of classroom			
Length of my table			
Around a quoit			
Depth of a sink			
Choose something else to measure			

(C) 1 Choose 2 things to measure in metres. Estimate and then measure.
What was the difference in length between them?
Write your answers on the back of this sheet.

2 Choose 2 things to measure in centimetres. Estimate and then measure. What was the difference in length between them?

Notes/date:

Cambridge Mathematics Direct 2 © Cambridge University Press 2002   M1.5

Doggy mixtures

Robot factory

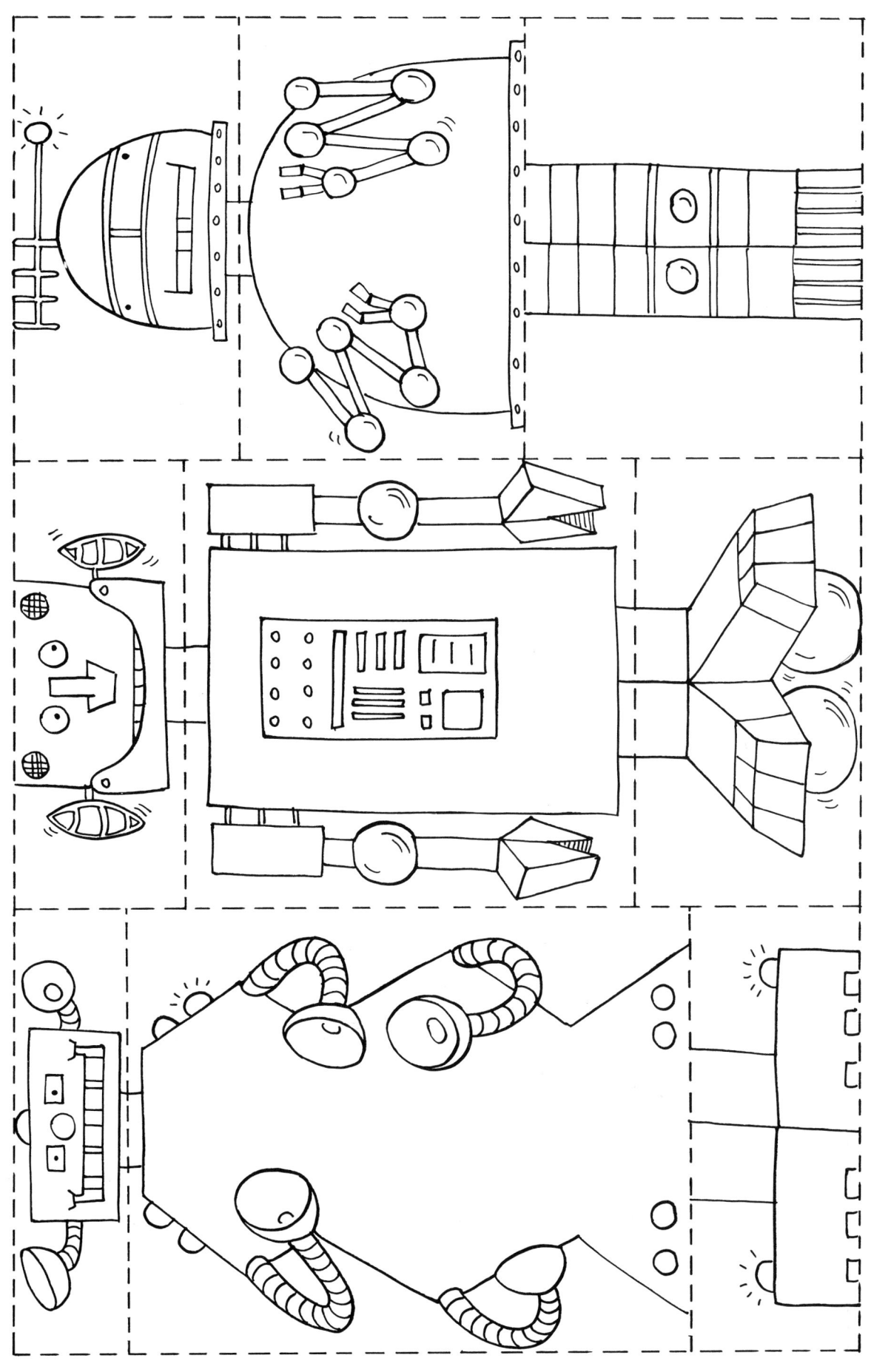

Making robots

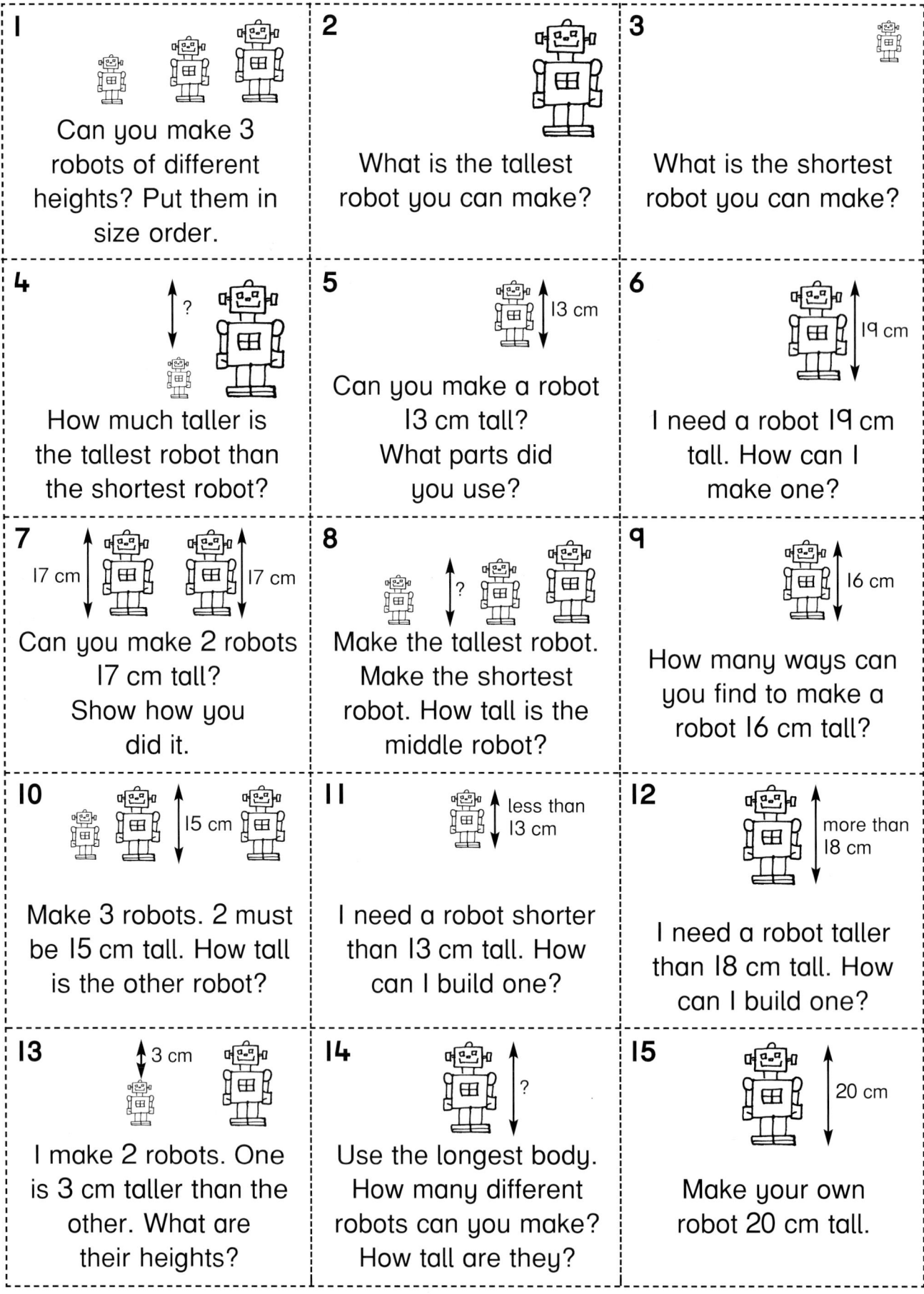

1 Can you make 3 robots of different heights? Put them in size order.

2 What is the tallest robot you can make?

3 What is the shortest robot you can make?

4 How much taller is the tallest robot than the shortest robot?

5 13 cm — Can you make a robot 13 cm tall? What parts did you use?

6 19 cm — I need a robot 19 cm tall. How can I make one?

7 17 cm | 17 cm — Can you make 2 robots 17 cm tall? Show how you did it.

8 ? — Make the tallest robot. Make the shortest robot. How tall is the middle robot?

9 16 cm — How many ways can you find to make a robot 16 cm tall?

10 15 cm — Make 3 robots. 2 must be 15 cm tall. How tall is the other robot?

11 less than 13 cm — I need a robot shorter than 13 cm tall. How can I build one?

12 more than 18 cm — I need a robot taller than 18 cm tall. How can I build one?

13 3 cm — I make 2 robots. One is 3 cm taller than the other. What are their heights?

14 ? — Use the longest body. How many different robots can you make? How tall are they?

15 20 cm — Make your own robot 20 cm tall.

Building a wall

Name:

Cut out the bricks below to help you. Each brick is 2 cm tall and 5 cm long.

(B) 1   What different walls can you make using all 18 bricks?
Write a multiplication sentence
for each one, e.g. 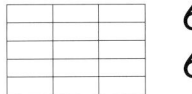   6 rows of 3 is 18
$6 \times 3 = 18$

(C) 2   A wall is 9 bricks long. How long is it in centimetres?
Write a multiplication sentence. $\boxed{9} \times \boxed{}$ cm $= \boxed{}$ cm

3   A wall is 5 bricks long.
How long is it in centimetres?
Write a multiplication sentence. $\boxed{} \times \boxed{}$ cm $= \boxed{}$ cm

4   I make a wall 8 cm tall. How many bricks tall is it?
Write a division sentence. $\boxed{}$ cm $\div \boxed{}$ cm $= \boxed{}$

5   How many bricks would you need to make a wall 30 cm long
and 6 cm high? Draw the wall.

Paving blocks 1

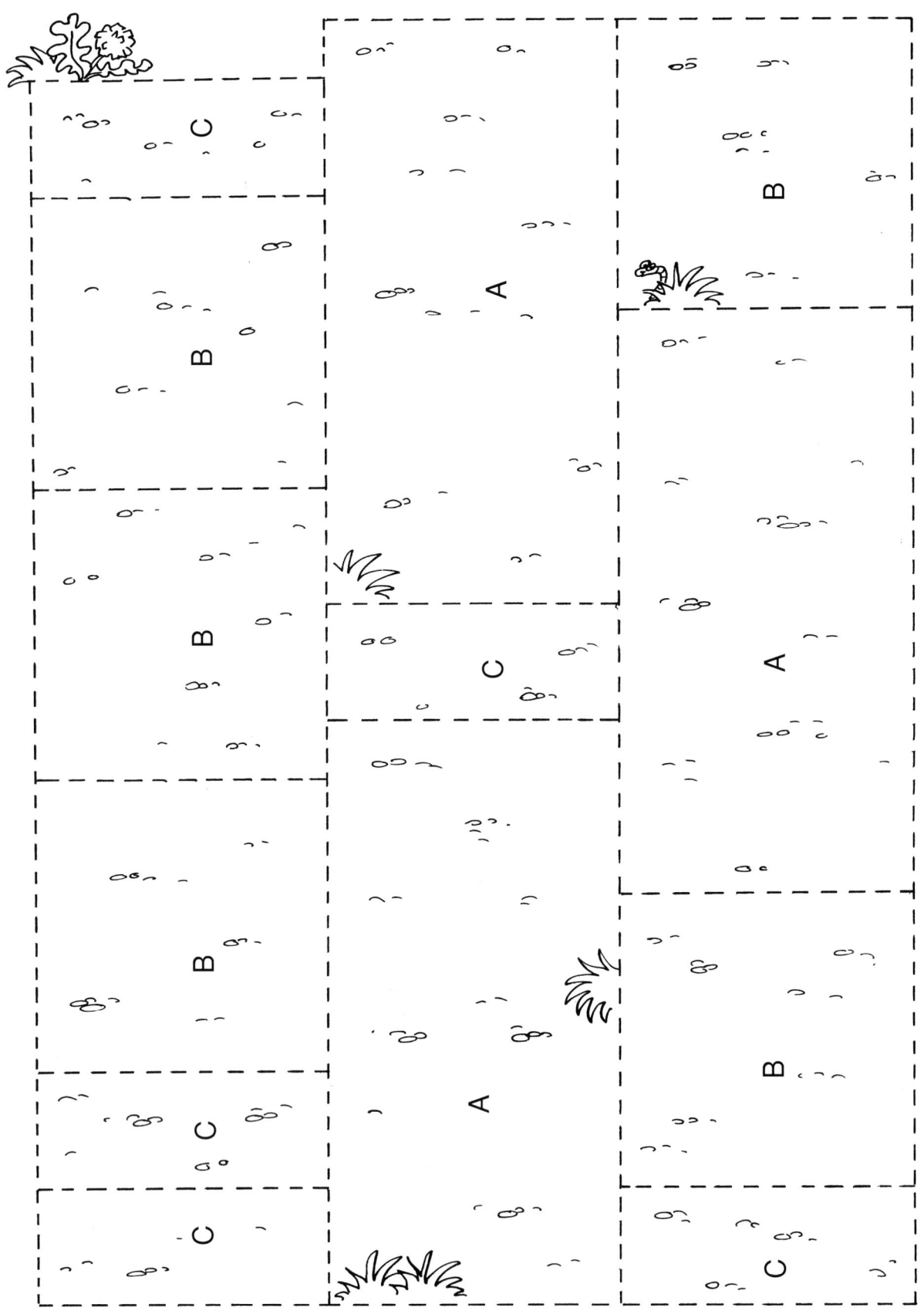

Paving blocks 2

1 Make a path 5 cm wide and 30 cm long. What blocks did you use?

|← 30 cm →|

2 Make a path using 3 B blocks. How long is it?

| B | B | B |

3

| A | B | B | C | C |

Make this path. How long is it?

4 What is the longest path you can make using just the A blocks on the sheet?

5 What is the longest path you can make using only the C blocks on the sheet?

6 Use all the blocks. How long is the path you make?

7 What is the shortest path you can make using 6 blocks?

8 Make these 2 paths:

| A | A | A |

| B | B | B | C | C |

Which path is longer? How much longer?

9 I use 3 B blocks to make a path and 3 C blocks to make another path. Which path is shorter? How much shorter?

10 How many A blocks would you need to make a path 50 cm long and 5 cm wide? Have you got enough?

11 How many more B blocks than A blocks would you need to make a path 30 cm long and 5 cm wide?

12 You use 3 A blocks. What other blocks could you use to make a path 40 cm long and 5 cm wide?

13 Can you make paths 30 cm long using 3 blocks, 4 blocks, 5 blocks, 6 blocks, 9 blocks?

14 A path is 10 cm long and 5 cm wide. How many more C blocks are needed to make it 20 cm long?

15 You have 4 B blocks. What other blocks would you need to make a path 27 cm long and 10 cm wide?

Heavier or lighter?

Name:

(A) Circle the heavier object.	Tick the lighter object.

1

4

2

5

3

6

Notes/date:

Cambridge Mathematics Direct 2 © Cambridge University Press 2002

M3.1

(14)

Fruit and vegetables

Name:

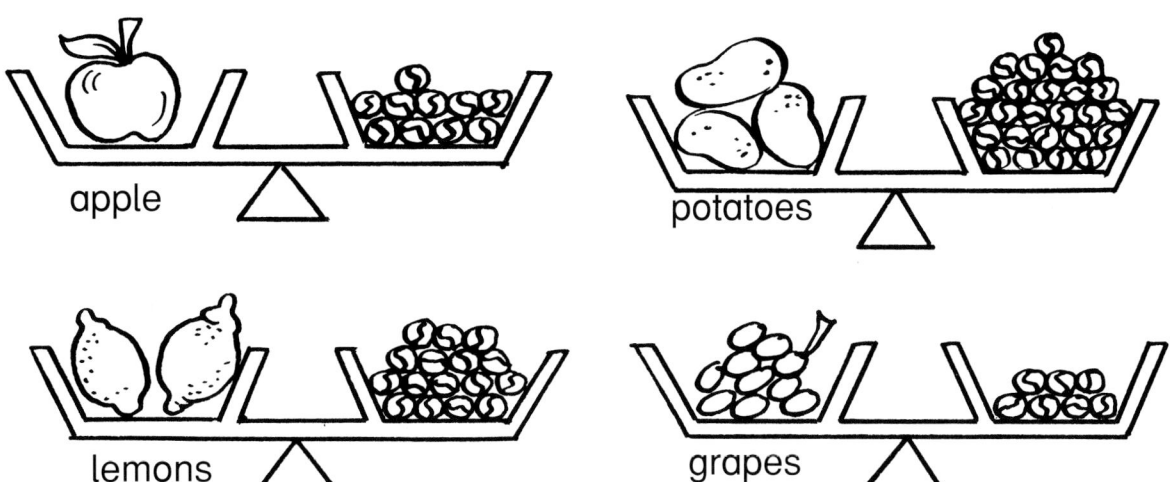

apple   potatoes

lemons   grapes

carrots

C **1** Put the fruit and vegetables in order from lightest to heaviest.

2 _____ weigh more than 3 carrots.

3 _____ and _____ are lighter than 2 lemons.

4 _____ and _____ are heavier than 2 lemons.

5 _____ weigh less than an apple.

6 How many apples weigh the same as 3 carrots? ☐

7 How many potatoes will weigh about the same as 50 marbles? ☐

How heavy?

(★) Estimate and draw.

heavier than I kg	lighter than I kg

What do you think is heaviest? _____

What do you think is lightest? _____

How many kilograms?

(A)

Object	My estimate	Actual weight

The _____ was the lightest. It weighed about ⬚ kg.

The _____ was the heaviest. It weighed about ⬚ kg.

What weighs about 1 kg?

(★) **1**   Circle the balance that shows about 1 kg.

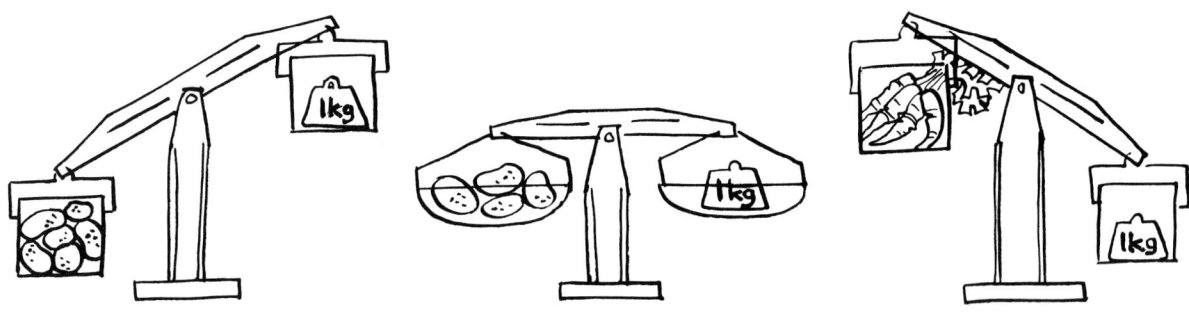

2   Can you estimate?

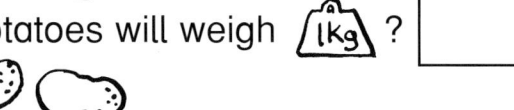

a   How many potatoes will weigh  ? ☐

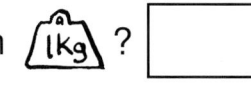

b   How many carrots will weigh 1kg ? ☐

c   How many bean bags will weigh 1kg ? ☐

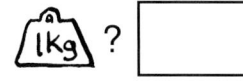

Notes/date:

Cambridge Mathematics Direct 2 © Cambridge University Press 2002   M3.3   (18)

What weighs about...?

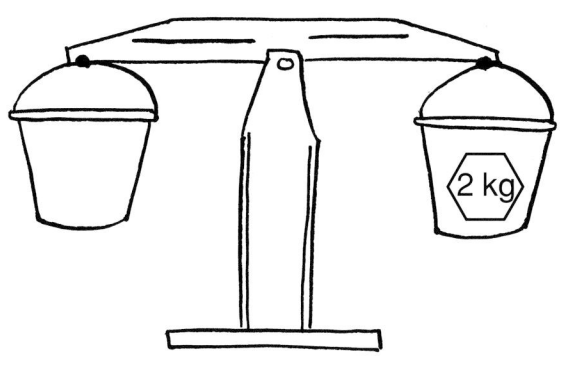

_____ weighs about 2 kg.

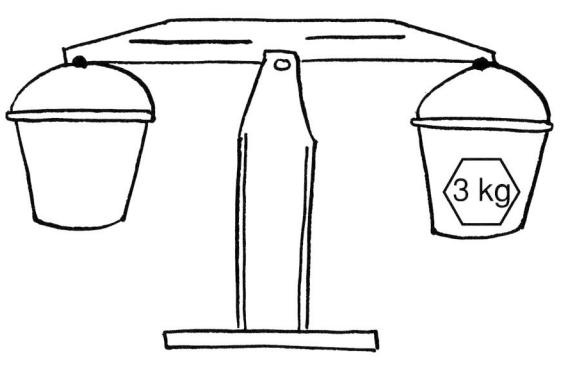

_____ weighs about 3 kg.

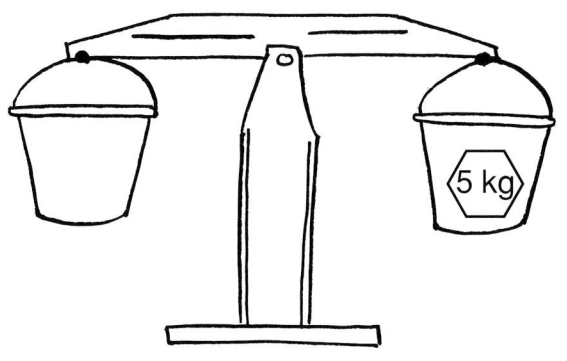

_____ weighs about 5 kg.

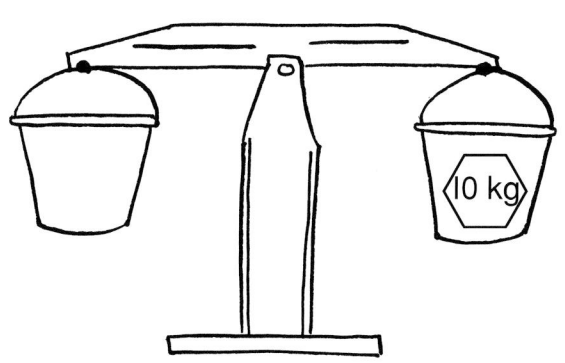

_____ weighs about 10 kg.

Notes/date:

Reading a scale

Name:

(A)   How much does it weigh?

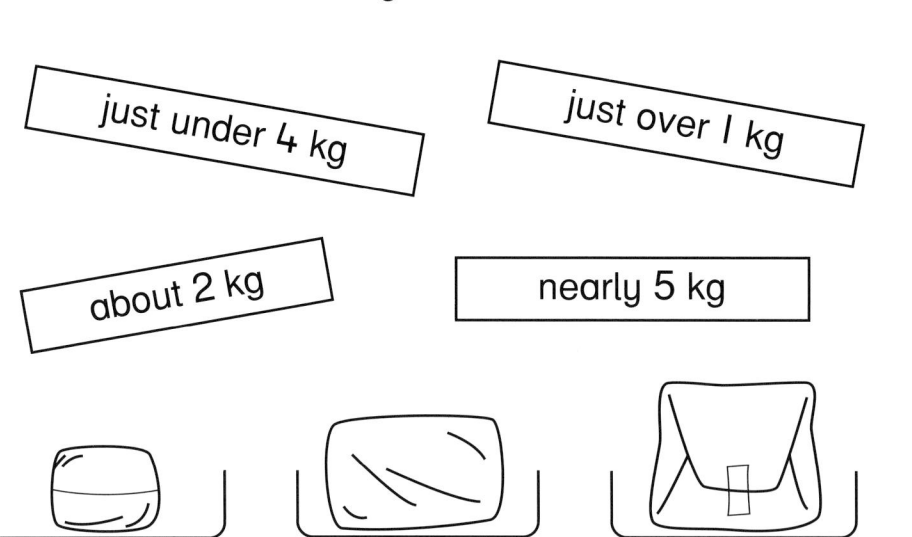

just under 4 kg

just over 1 kg

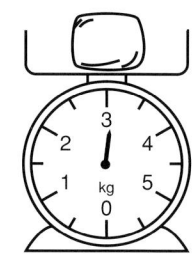

about 2 kg

nearly 5 kg

about 3 kg

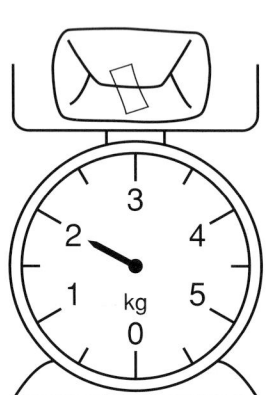

_____   _____   _____   _____

Draw an arrow to show the weight.

just over 4 kg

nearly 1 kg

just under 2 kg

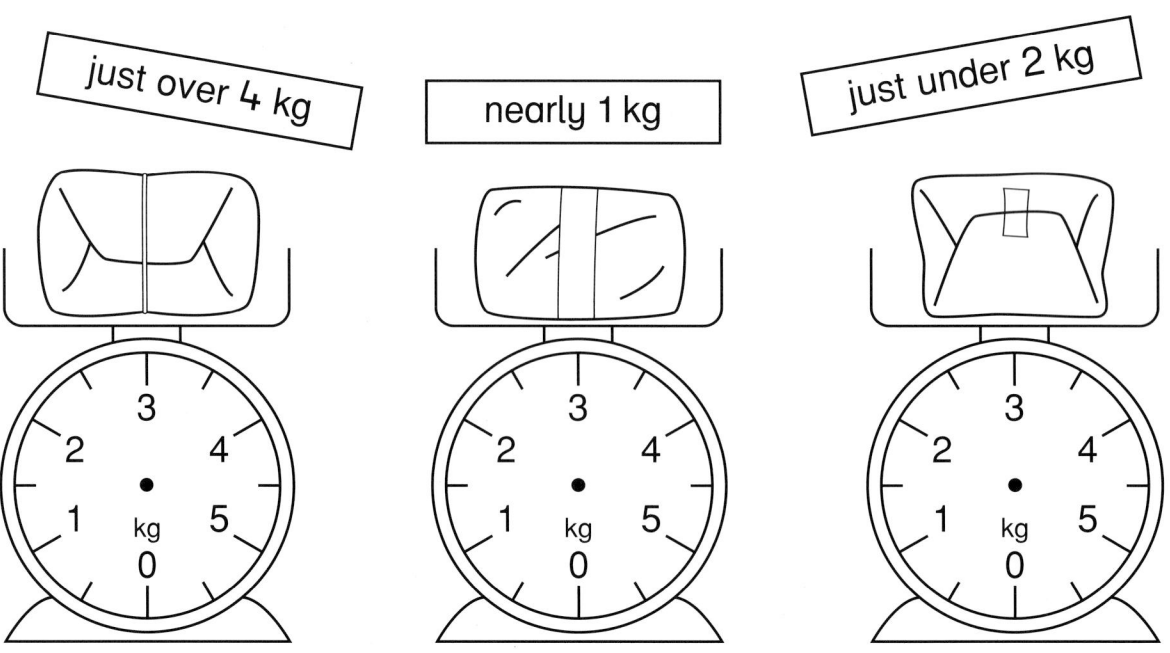

1000 g = 1 kilogram

Name:

(A) 1 Find things to weigh and draw.

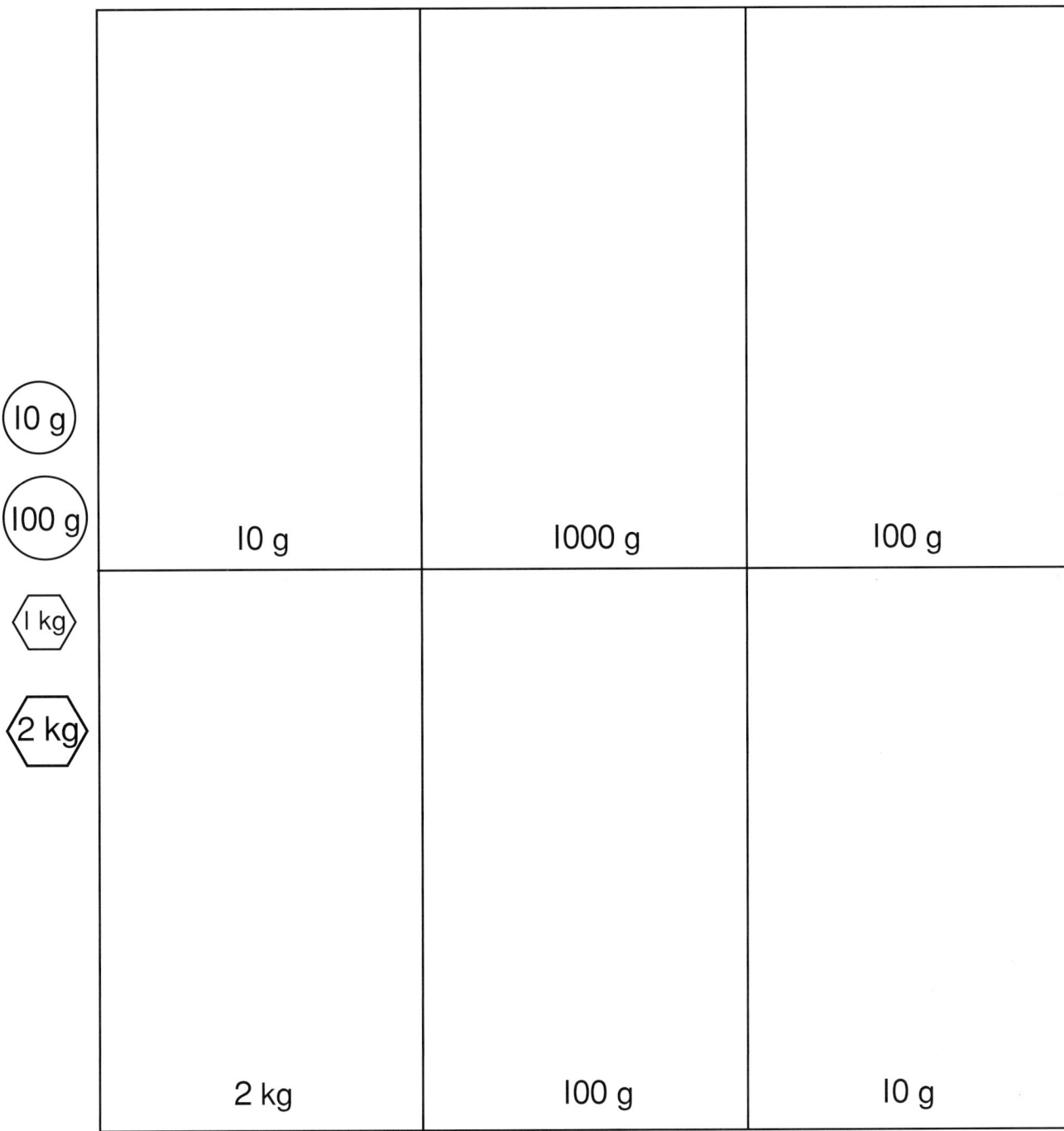

10 g	1000 g	100 g
2 kg	100 g	10 g

2 How many grams or kilograms?

⟨1 kg⟩ = ☐ g   4000 g = ☐ kg

⟨1 kg⟩ ⟨1 kg⟩ ⟨1 kg⟩ = ☐ g   2000 g = ☐ kg

⟨1 kg⟩ ⟨1 kg⟩ ⟨1 kg⟩ ⟨1 kg⟩ ⟨1 kg⟩ = ☐ g   7000 g = ☐ kg

Notes/date:

The post

Name:

1 kg = 1000 g	$\frac{1}{2}$ kg = 500 g

1 The postman needs to know how many grams his parcels weigh.

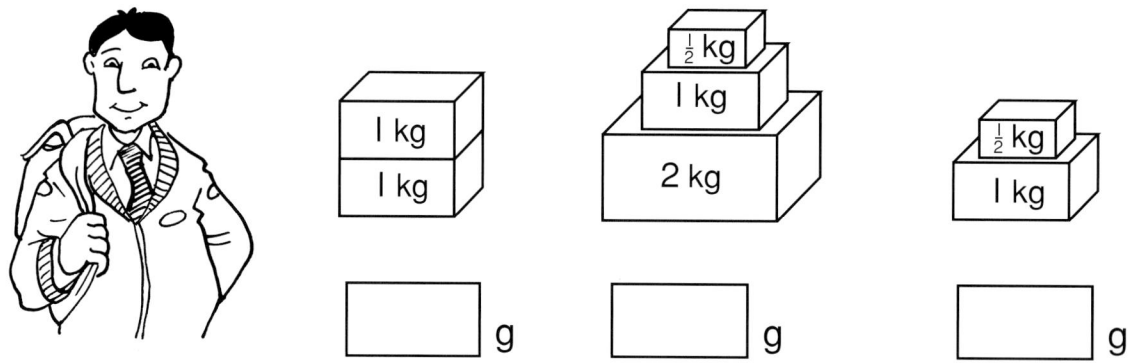

☐ g   ☐ g   ☐ g

2 The postlady needs to know how many kilograms her parcels weigh.

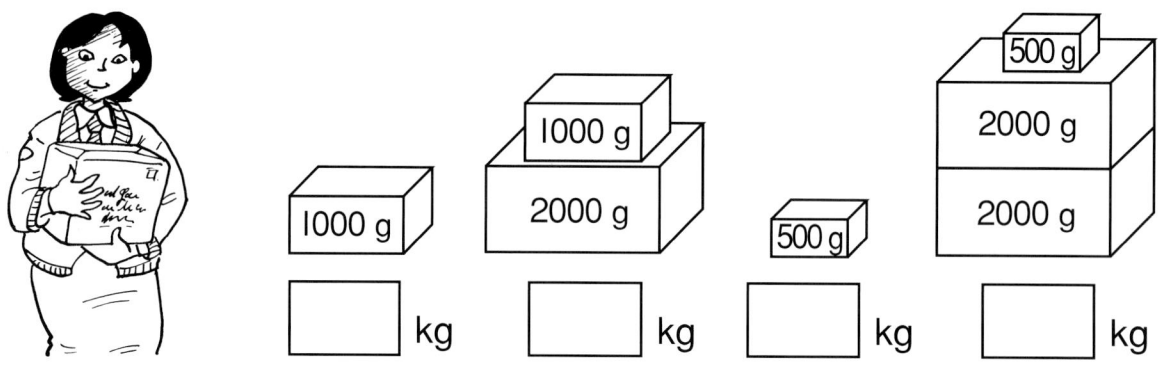

☐ kg   ☐ kg   ☐ kg   ☐ kg

3 Write your own weights on these parcels, in kilograms and grams.

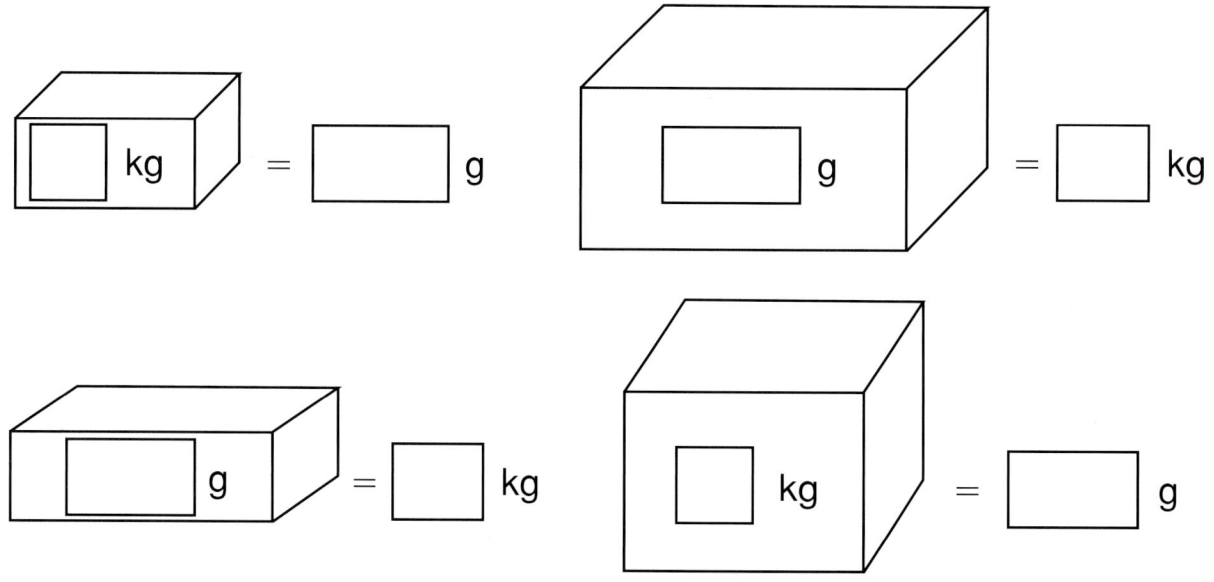

Notes/date:

Greengrocer's shop 1

Greengrocer's shop 2

1

What is in the heaviest sack? How heavy is it?

2

What is in the lightest sack? How much does it weigh?

3

Put the three sacks in order, lightest first.

4

How much do all 3 bags of fruit weigh altogether?

5

How much would 3 sacks of carrots weigh altogether?

6

Sara buys 2 large bags of carrots and 3 bags of oranges. How heavy is this altogether?

7

Which is heavier, a bag of apples or a bag of oranges? How much heavier?

8

Which is lighter, a sack of potatoes or a large bag of turnips? How much lighter?

9

I need 47 kg of carrots to feed my horse. How many sacks and bags do I need to buy?

10

How many different ways can you find to buy 5 kg of fruit?

11

What is the difference in weight between a bunch of bananas and a lemon?

12

Which weighs more, 2 large bags of turnips or 6 small bags of turnips? How much more?

13

Our school is given a 10 kg box of different fruits for a raffle prize. What might be in the box?

14

How much would a bunch of bananas, a grapefruit and 3 lemons weigh altogether?
1 kg = 1000 g

15

How many bunches of bananas would make 2 kg?

1 kg = 1000 g

Bridges 1

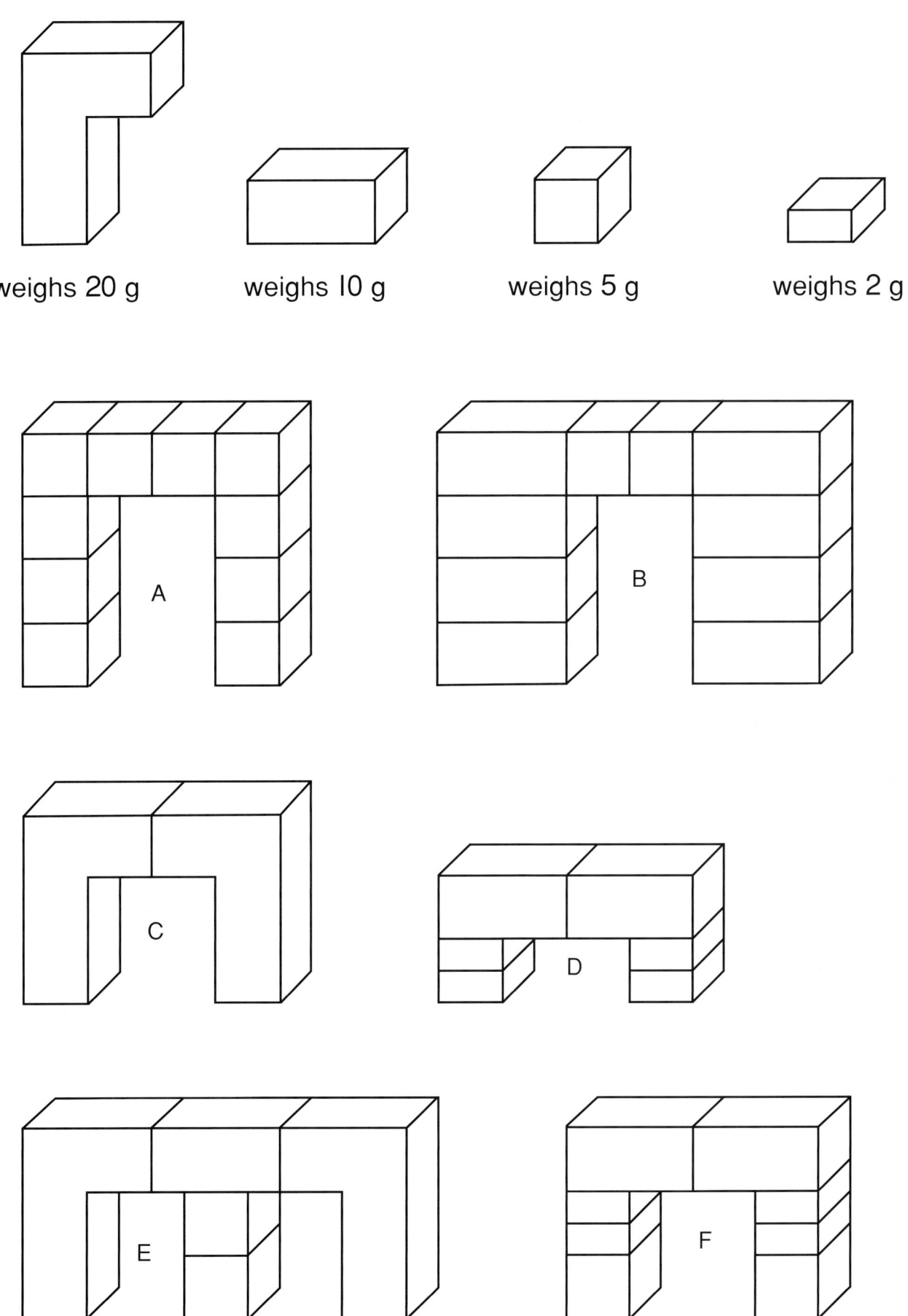

weighs 20 g   weighs 10 g   weighs 5 g   weighs 2 g

Bridges 2

1 Bridge A uses ten 5 g pieces. How heavy is it?	**2** Bridge D uses two 10 g pieces and four 2 g pieces. How heavy is it?	**3** How heavy is bridge C?
4 Draw your own bridge. How heavy is it?	**5** What is the difference in weight between bridge A and bridge B?	**6** How much heavier is bridge B than bridge C?
7 Which is lighter, bridge A or bridge C?	**8** How much does bridge F weigh?	**9** Try to design a bridge that weighs 70 g.
10 The supports of bridge D are too short. Make them twice as tall. How heavy is bridge D now?	**11** Make bridge D twice as wide. How heavy is it now? (**Tip:** it will need more supports.)	**12** What is the difference in weight between bridge C and bridge E?
13 Try to design a bridge that weighs 44 g. Find another way.	**14** To make another tunnel, bridge C is built next to bridge E. What is their total weight?	**15** Try to design a bridge twice as heavy as bridge C.

Who am I?

Name:

C   Circle the letter for the pot (or pots) that matches the sentence:

A B C (D) E   none of them

1   I hold about 5 cups.

    I am A B C D E   none of them

2   I hold more than 6 cups.

    I am A B C D E   none of them

3   I hold less than 4 cups.

    I am A B C D E   none of them

4   I have the greatest capacity.

    I am A B C D E

5   I have the least capacity.

    I am A B C D E

6   I hold more than 8 cups, but less than 12 cups.

    I am A B C D E   none of them

Notes/date:

Cambridge Mathematics Direct 2 © Cambridge University Press 2002   M5.1

Measuring bottles

Name:

★ **1** How much does each container hold?
Use your bottle to find out.

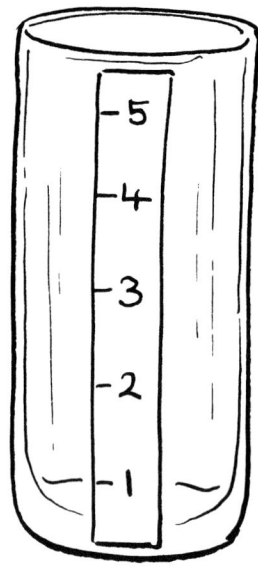

☐ cups   ☐ cups   ☐ cups

Ⓐ **2** Which holds the most? ☐

3 Which holds the least? ☐

4 Put the 3 containers in order, from smallest to
greatest capacity.

☐   ☐   ☐

Measuring in litres

B How many litres do these containers hold?

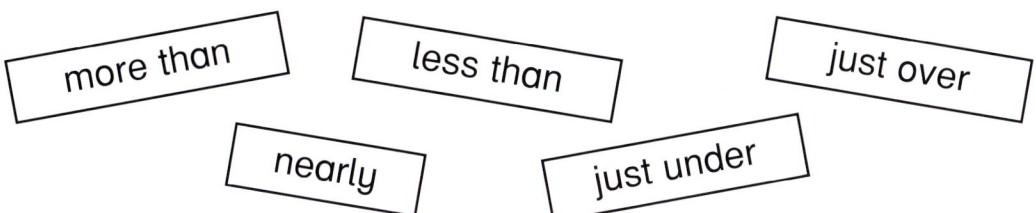

more than   less than   just over   nearly   just under

1 estimate _____ litres capacity _____ litres	**2** 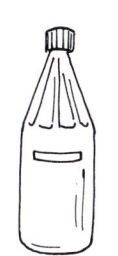 estimate _____ litres capacity _____ litres
3  estimate _____ litres capacity _____ litres	**4**  estimate _____ litres capacity _____ litres
5  estimate _____ litres capacity _____ litres	**6** Choose and draw a container. estimate _____ litres capacity _____ litres

Notes/date:

About a litre

Name:

under a litre
about a litre
over a litre

almost a litre

nearly a litre

just over a litre

just under a litre

just more than a litre

about a litre

just less than a litre

Notes/date:

Reading a scale

Name:

Ⓑ

nearly   just over   almost   just less than

just more than   exactly   just under

How much is in these containers?

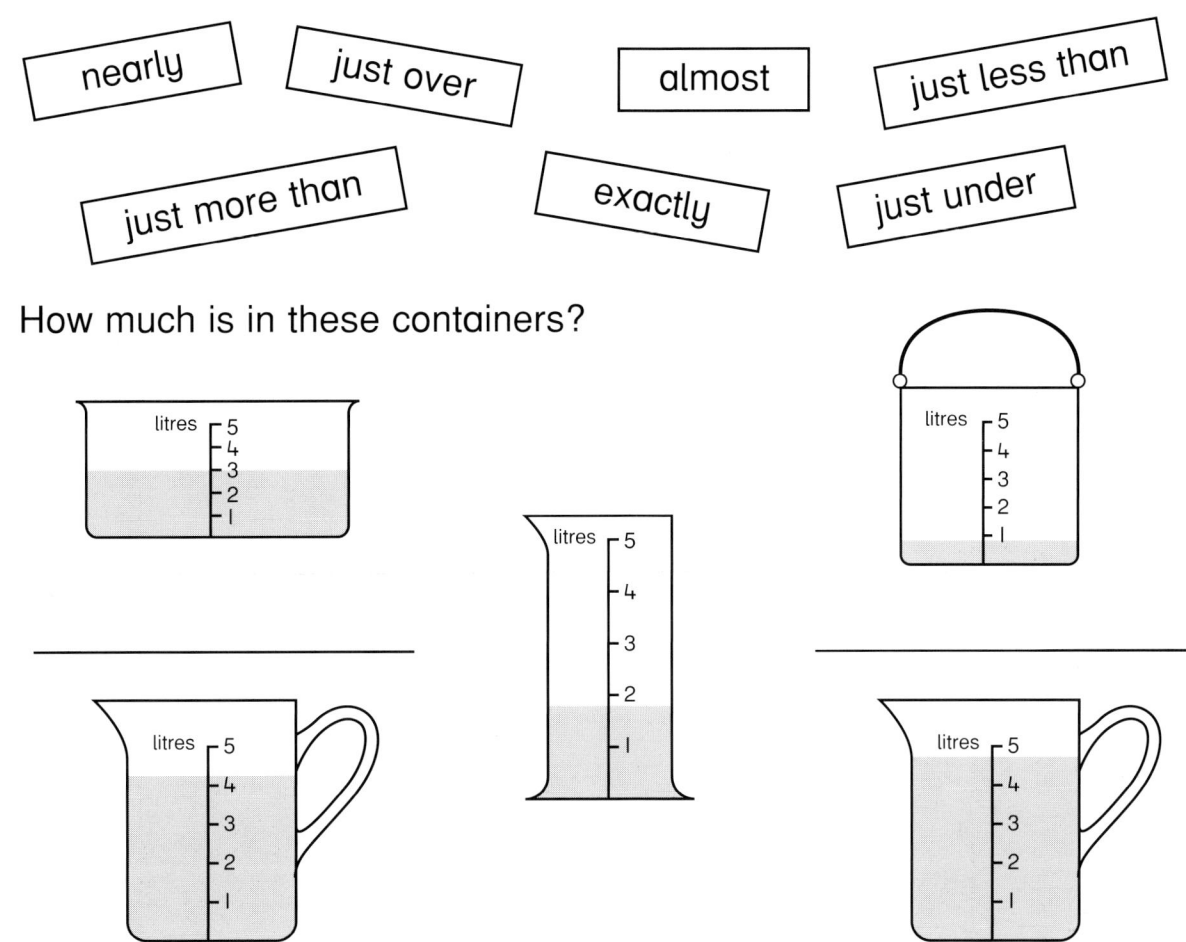

_____   _____

_____   _____

Draw a line in each container to show the level of juice.

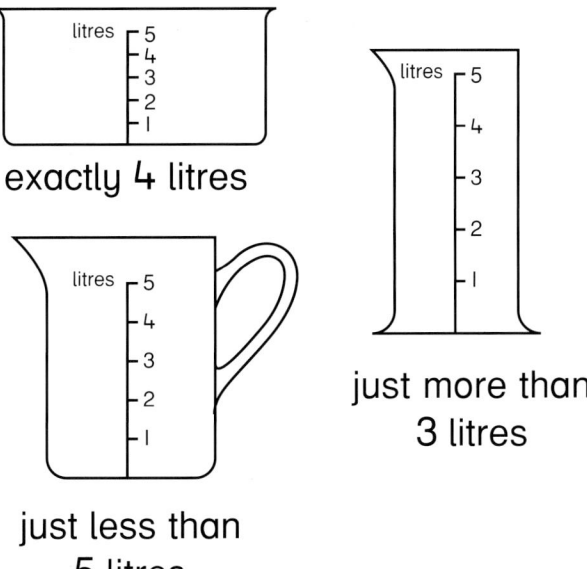

exactly 4 litres

just less than
5 litres

just more than
3 litres

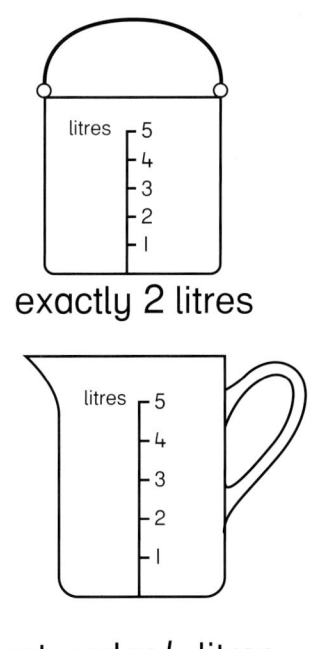

exactly 2 litres

just under 4 litres

Measuring capacity

Name:

B How much?

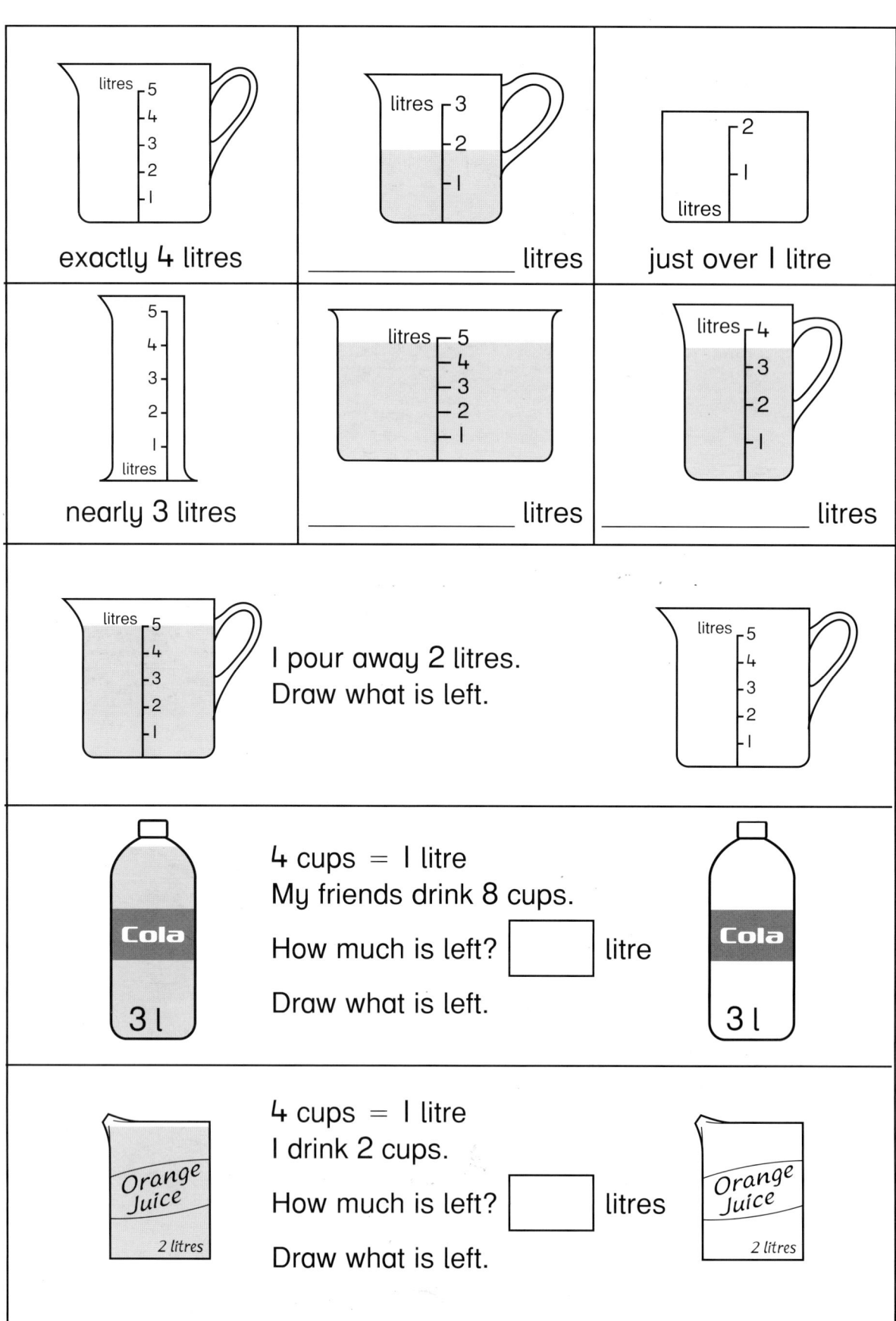

exactly 4 litres

_____ litres

just over 1 litre

nearly 3 litres

_____ litres

_____ litres

I pour away 2 litres.
Draw what is left.

4 cups = 1 litre
My friends drink 8 cups.

How much is left? ☐ litre

Draw what is left.

Cola 3 l

4 cups = 1 litre
I drink 2 cups.

How much is left? ☐ litres

Draw what is left.

Orange Juice 2 litres

Notes/date:

Cambridge Mathematics Direct 2 © Cambridge University Press 2002

M5.5

32

Water at home

How much water does it use?

Washing up	22 litres	Washing hands and face	7 litres
Washing machine	78 litres	Bath	83 litres
Boiling a kettle	I litre	Shower	43 litres
Flushing the toilet	8 litres	Washing the car	30 litres
Cleaning teeth	2 litres	Filling the paddling pool	50 litres

Cambridge Mathematics Direct 2 © Cambridge University Press 2002   M6.1

1

Which 3 activities use 17 litres of water altogether?

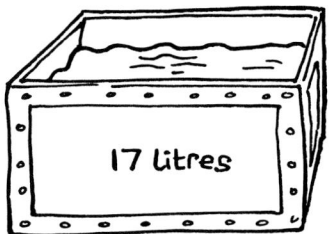

17 litres

2

How much water is used in total by washing the car and having a shower?

3

How much more water is used filling the paddling pool than having a shower?

4

How much more water is used by the washing machine than by flushing the toilet?

5

Cal washes his car twice a month. How much water does he use for this each month?

6

Two showers use more than one bath. How much more?

7

Which 2 things use exactly 100 litres of water in total?

100 litres

8

What is the difference between the amounts of water used by the bath and the washing machine?

Water at home question cards

Cambridge Mathematics Direct 2 © Cambridge University Press 2002   M6.1

The ice cream seller

One litre  makes 10 single cones

or 5 double cones  .

(B) **1** Complete the table with the number of double or single cones or the amount of litres sold.

Ice cream sold		Vanilla	Chocolate
Thursday		20 single cones ☐ litres	40 single cones ☐ litres
		15 double cones ☐ litres	30 double cones ☐ litres
Friday		☐ single cones 5 litres	☐ single cones 3 litres
		☐ double cones 4 litres	☐ double cones 2 litres

(C) **2** 6 litres of vanilla ice cream was sold on Saturday.

This was sold as 40 single cones and ☐ double cones.

3 In what ways could 3 litres of chocolate ice cream have been sold?

Example: 20 = 2 litres

5  = 1 litre

2 litres + 1 litre = 3 litres

Find another way.

Months and seasons

Name:

(A)

June	August	September	April
November	~~January~~	October	~~December~~
May	March	July	February

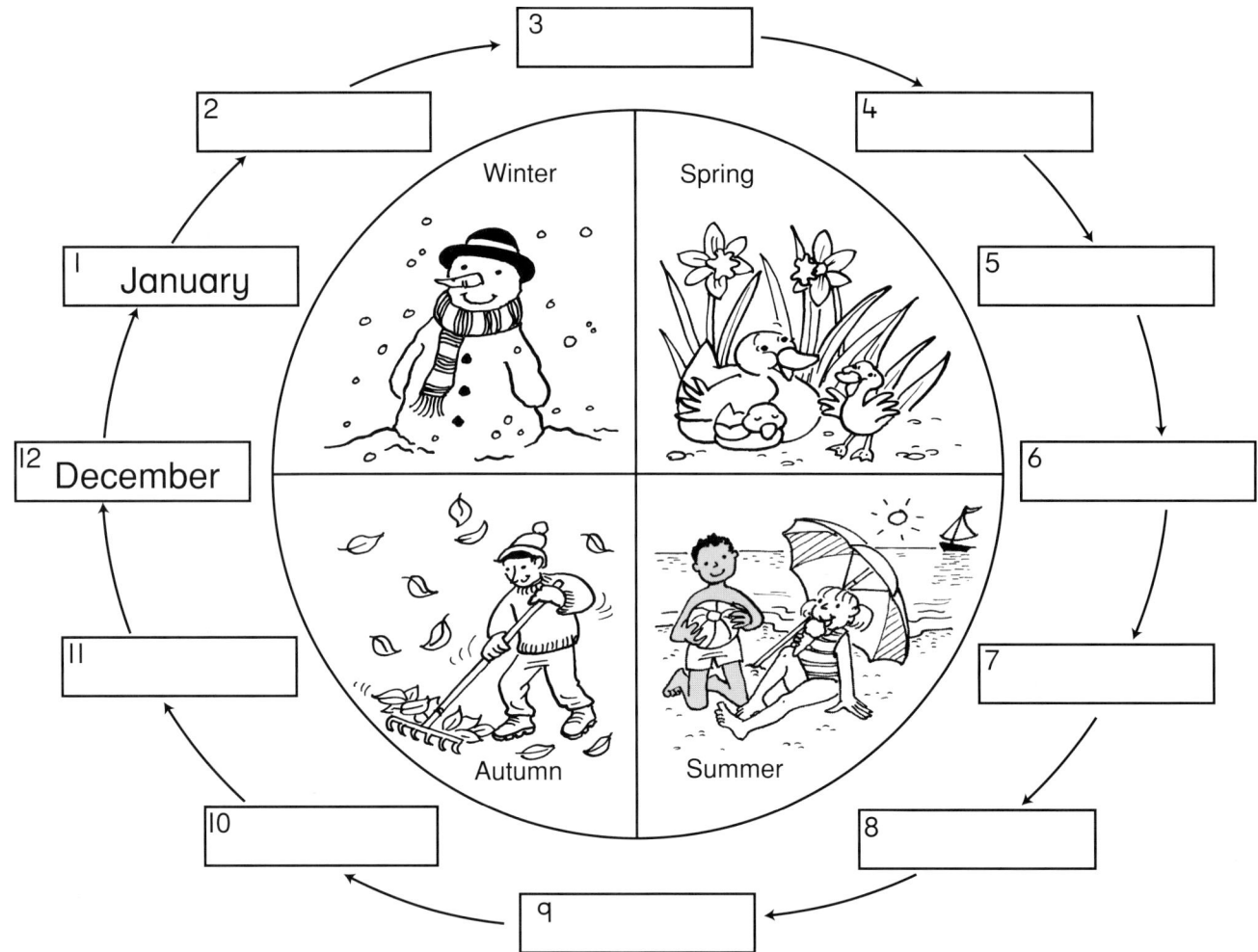

(B)   Special dates:

My birthday _____ of _____

_____ of _____

_____ of _____

April	September	August	June
December	October	January	November
February	July	March	May

Months of the year cards

Months of the year

Name:

Write the month that comes **after** each month.
Draw a picture to show the season.

January   _February_

March   _____

May   _____

July   _____

December   _____

November   _____

Write the month that comes **before** each month.
Draw a picture to show the season.

March   April

_____   October

_____   June

_____   February

_____   August

_____   January

Cambridge Mathematics Direct 2 © Cambridge University Press 2002   M7.1

O'clock and half past

Name:

Make the digital and analogue clocks say the same time.

Notes/date:

Cambridge Mathematics Direct 2 © Cambridge University Press 2002   M7.2

Cinderella

Name:

C Show the times when the show starts and finishes on the analogue clocks.

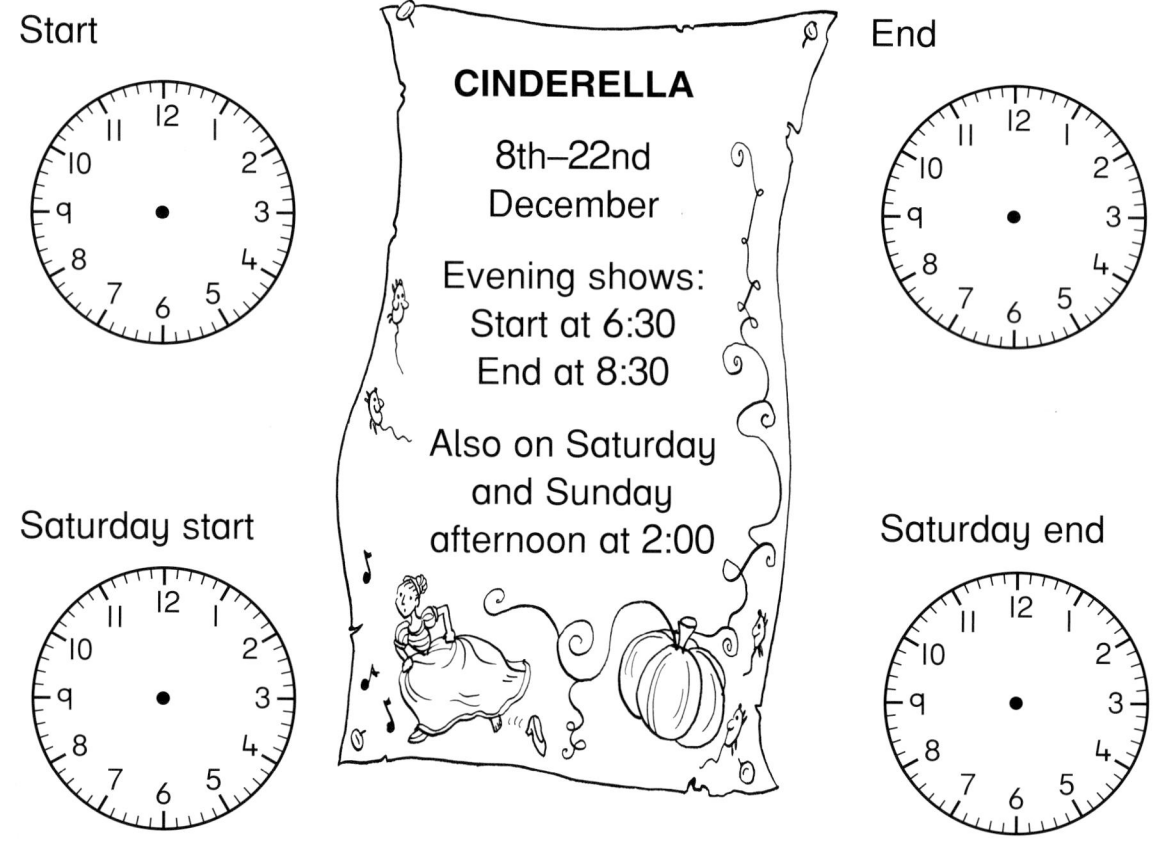

Start

End

Saturday start

Saturday end

CINDERELLA

8th–22nd December

Evening shows:
Start at 6:30
End at 8:30

Also on Saturday and Sunday afternoon at 2:00

I How long does the pantomime last?

_____

2 When does the 2:00 performance on Saturday end?

_____

3 Kala arrives at 6:00.
How long does she have to wait until the show starts?

_____

4 Katie gets home at 9:30 after the evening show.
How long did it take her to get home?

_____

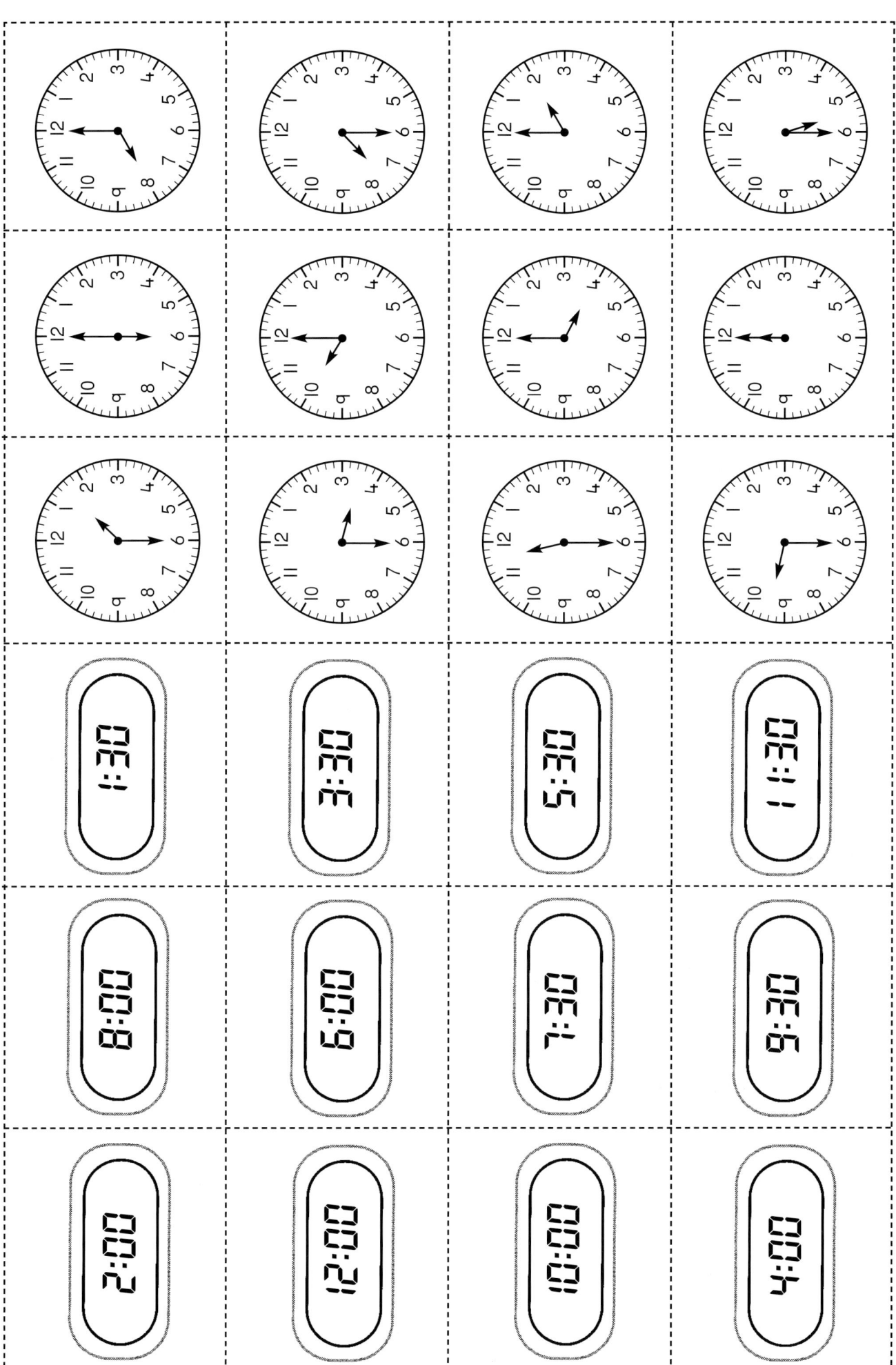

Quarter past

Name:

Show the times on the digital and analogue clocks.

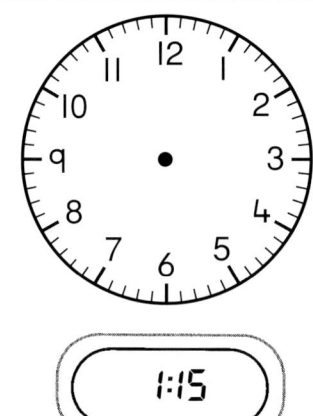 **1:15** The time is quarter past _____.		

The time is quarter past ____2____.

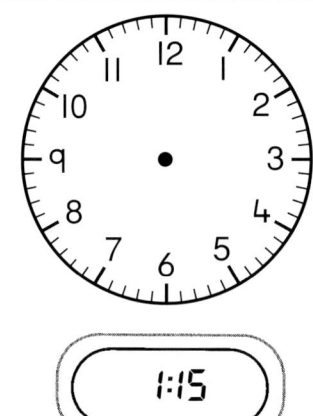

1:15

The time is quarter past _____.

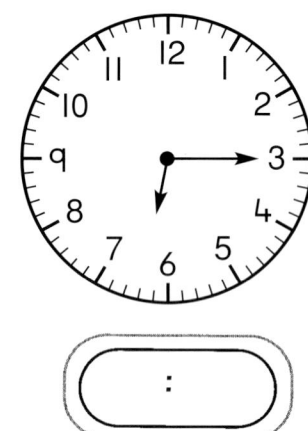

:

The time is quarter past _____.

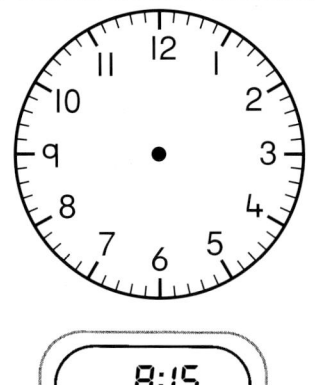

8:15

The time is quarter past _____.

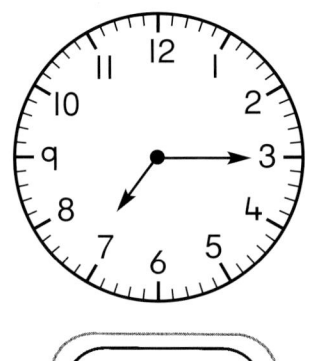

:

The time is quarter past _____.

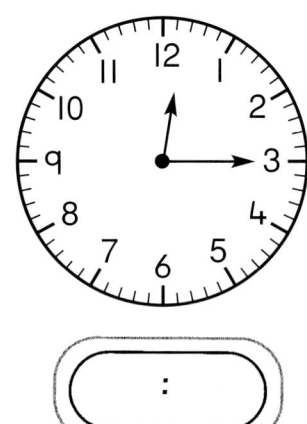

:

The time is quarter past _____.

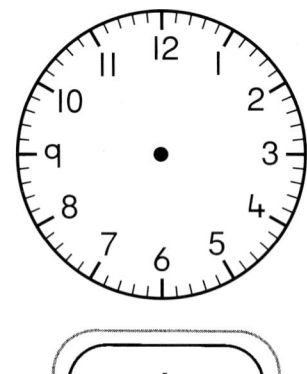

:

The time is quarter past 9.

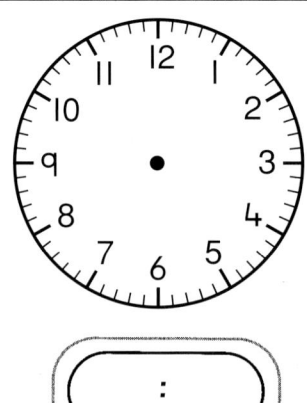

:

The time is quarter past 5.

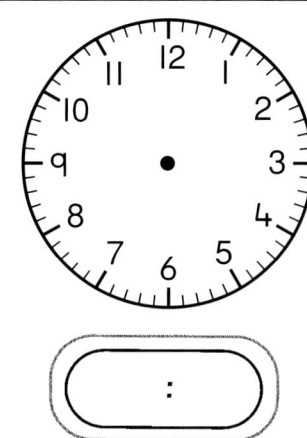

:

The time is quarter past 3.

Notes/date:

Cambridge Mathematics Direct 2 © Cambridge University Press 2002

15 minutes past 6		6:15	quarter past 6
15 minutes past 10		10:15	quarter past 10
quarter past 1		1:15	15 minutes past 1
quarter past 7		7:15	15 minutes past 7
15 minutes past 3		3:15	quarter past 3
quarter past 11		11:15	15 minutes past 11
15 minutes past 12		12:15	quarter past 12

Clock families

I hour later	I hour earlier
half an hour later	half an hour earlier
$\frac{1}{2}$ hour later	$\frac{1}{2}$ hour earlier
30 minutes later	30 minutes earlier
quarter of an hour later	quarter of an hour earlier
$\frac{1}{4}$ hour later	$\frac{1}{4}$ hour earlier
15 minutes later	15 minutes earlier

Earlier or later

Holiday flight to Majorca

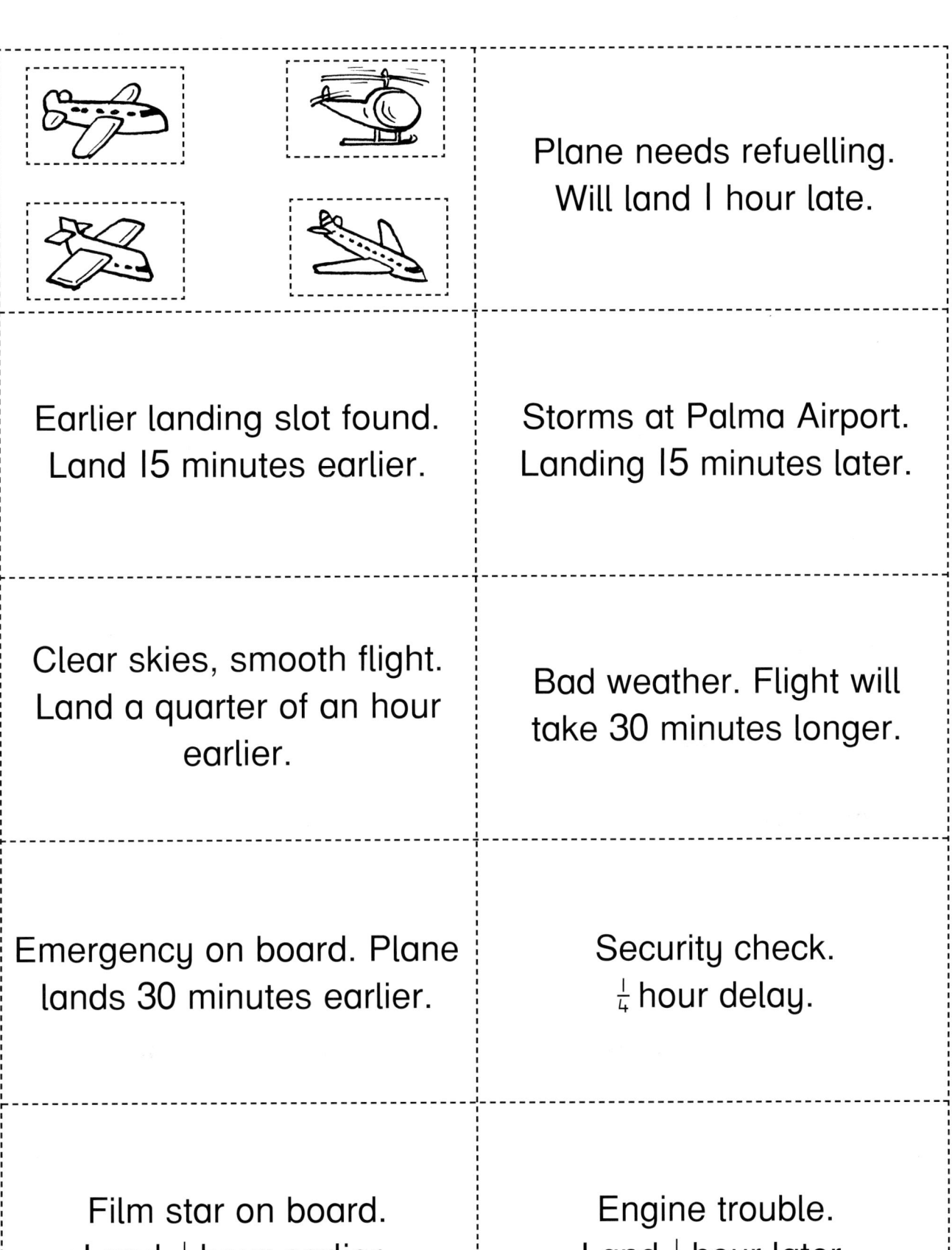

Plane needs refuelling.
Will land 1 hour late.

Earlier landing slot found.
Land 15 minutes earlier.

Storms at Palma Airport.
Landing 15 minutes later.

Clear skies, smooth flight.
Land a quarter of an hour
earlier.

Bad weather. Flight will
take 30 minutes longer.

Emergency on board. Plane
lands 30 minutes earlier.

Security check.
$\frac{1}{4}$ hour delay.

Film star on board.
Land $\frac{1}{2}$ hour earlier.

Engine trouble.
Land $\frac{1}{2}$ hour later.

Cambridge Mathematics Direct 2 © Cambridge University Press 2002   M7.4

Holiday flight to Majorca
(a game for 2–4 players)

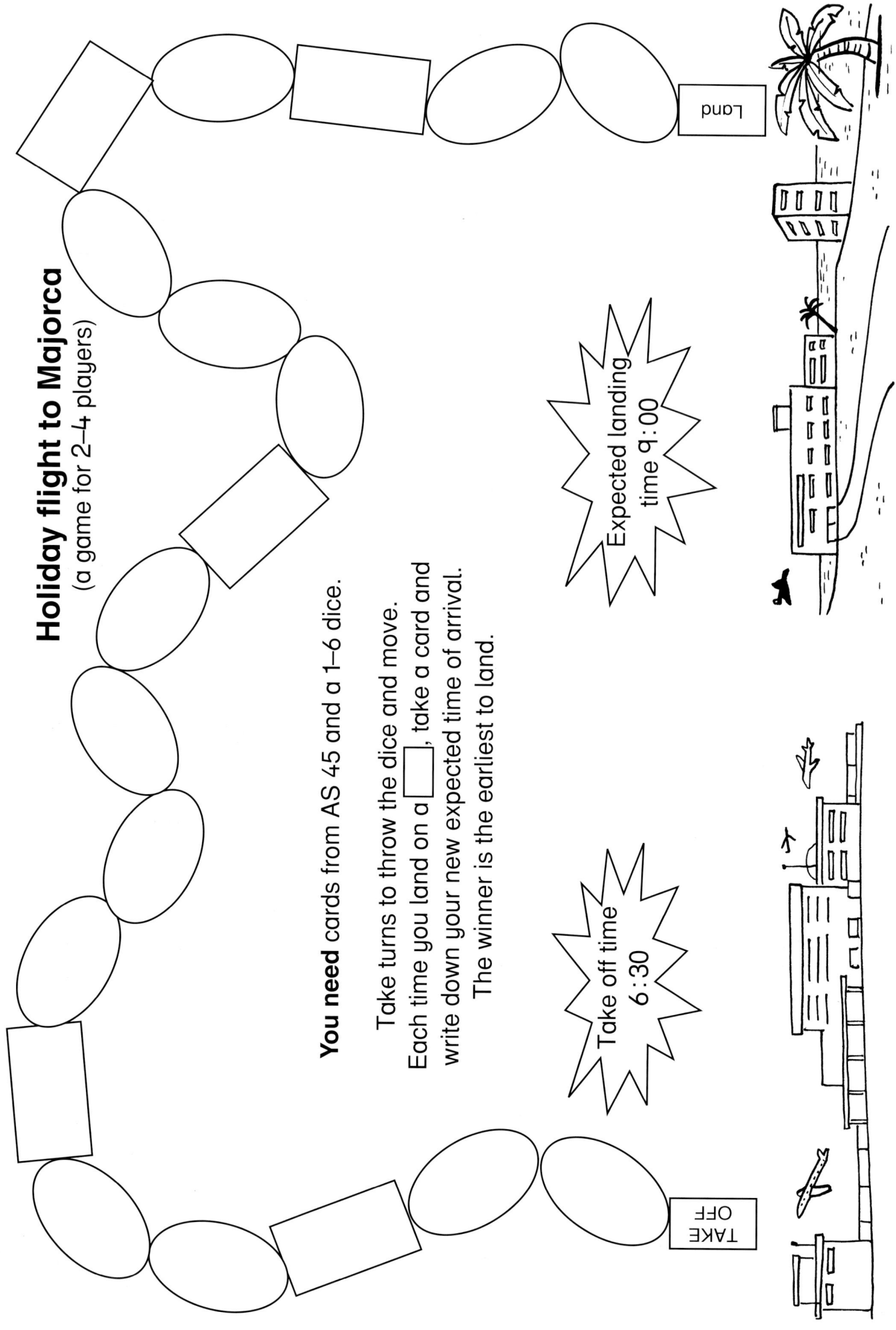

You need cards from AS 45 and a 1–6 dice.

Take turns to throw the dice and move.
Each time you land on a ☐, take a card and
write down your new expected time of arrival.
The winner is the earliest to land.

Expected landing
time 9:00

Take off time
6:30

Land

TAKE OFF

Party time

A How long is the party?

Party begins   Party ends

   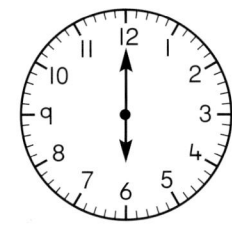   The party was ☐ hours long.

      The party was ☐ hours long.

      The party was ☐ hours long.

Party begins   Party ends

   The party was 3 hours long.
When did it end?
Draw the time.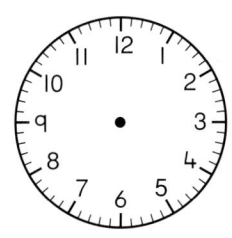

The party was 5 hours long.
When did it start?
Draw the time.

3:45   The party was 2 hours long.
When did it end?
Write the time.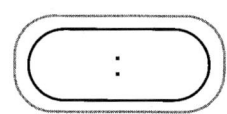

Notes/date:

Party began at 12:30.

Party ended at 4:30. How long did the party last?

Party lasted 1 hour. When did the party end?

Party began at 1:30.

Party ended at 5:30. How long did the party last?

Party lasted 2 hours. When did the party end?

Party began at 2:30.

Party ended at 6:30. How long did the party last?

Party lasted 3 hours. When did the party end?

Party began at 3:30.

Party ended at 7:30. How long did the party last?

Party lasted 4 hours. When did the party end?

Party game cards

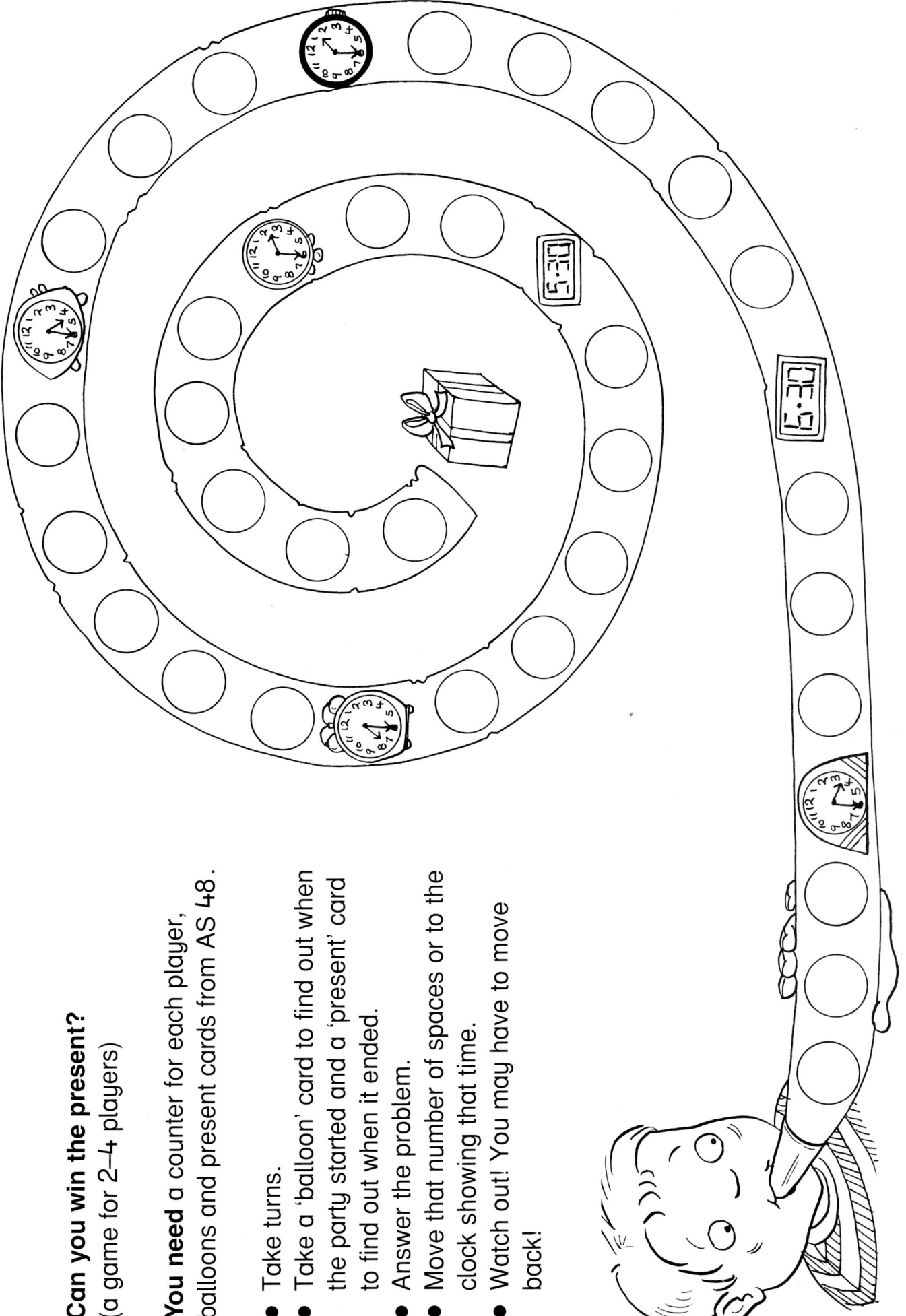

Can you win the present?
(a game for 2–4 players)

You need a counter for each player, balloons and present cards from AS 48.

- Take turns.
- Take a 'balloon' card to find out when the party started and a 'present' card to find out when it ended.
- Answer the problem.
- Move that number of spaces or to the clock showing that time.
- Watch out! You may have to move back!

Party game

Going swimming

Name:

How long did the children spend in the pool altogether?

	slide pool		big pool		How long altogether?

Tom $\frac{1}{2}$ an hour   + $\frac{1}{2}$ an hour   =   ☐ hour

Ann   1 hour   + $\frac{1}{2}$ an hour   =   ☐ hours

How long did the children spend in the big pool?

slide pool   big pool   How long altogether?

Bob   1 hour   +   ☐ hour   = 2 hours

Amit $\frac{1}{2}$ an hour   +   ☐ hours   = 2 hours

How long did the children spend in the slide pool?

slide pool   big pool   How long altogether?

Sophie   ☐ minutes   +   15 minutes   = 45 minutes

Sid   ☐ minutes   +   30 minutes   = 1 hour

Notes/date:

Cambridge Mathematics Direct 2 © Cambridge University Press 2002

M7.6

(50)

The swimming pool

Fill in the missing times.

 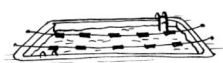

started	slide pool	big pool	finished
	I hour	$\frac{1}{2}$ hour	
	☐ hour	I hour	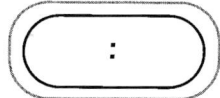
6:00	I5 minutes	30 minutes	:
:	I hour	30 minutes	5:00
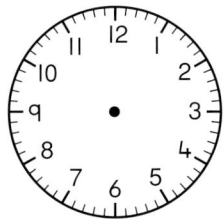	30 minutes	I hour	
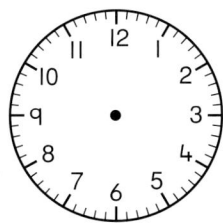	45 minutes	45 minutes	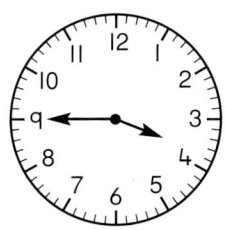

How many sides?

(A) **You need** coloured pencils.
Colour: 3-sided shapes yellow
4-sided shapes red
5-sided shapes blue.

Flat shapes

B I Fill in the boxes.

3 sides
3 corners

a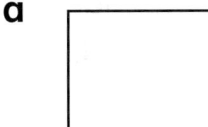
□ sides
□ corners

b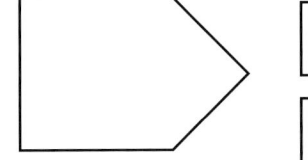
□ sides
□ corners

c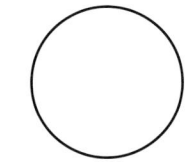
□ sides
□ corners

e Draw a shape of your own.

d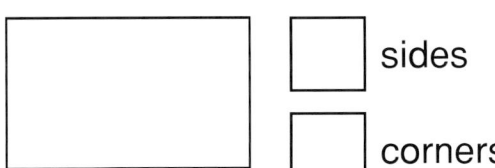
□ sides
□ corners

□ sides
□ corners

2 Match the shapes to their names.

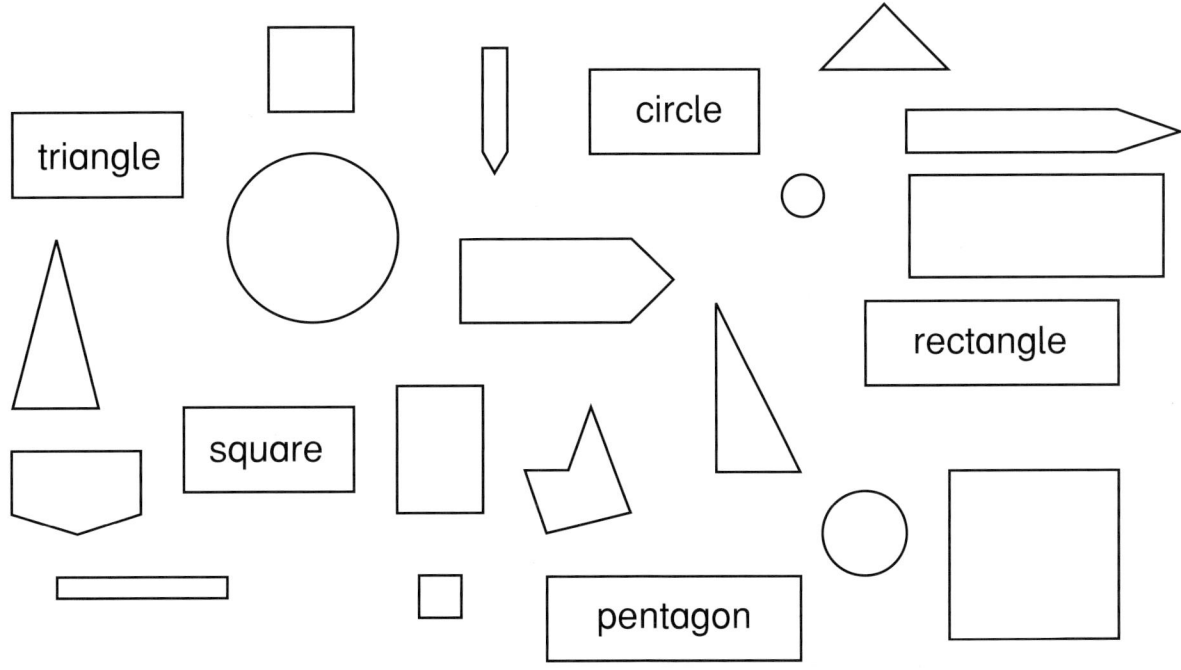

triangle   circle   square   rectangle   pentagon

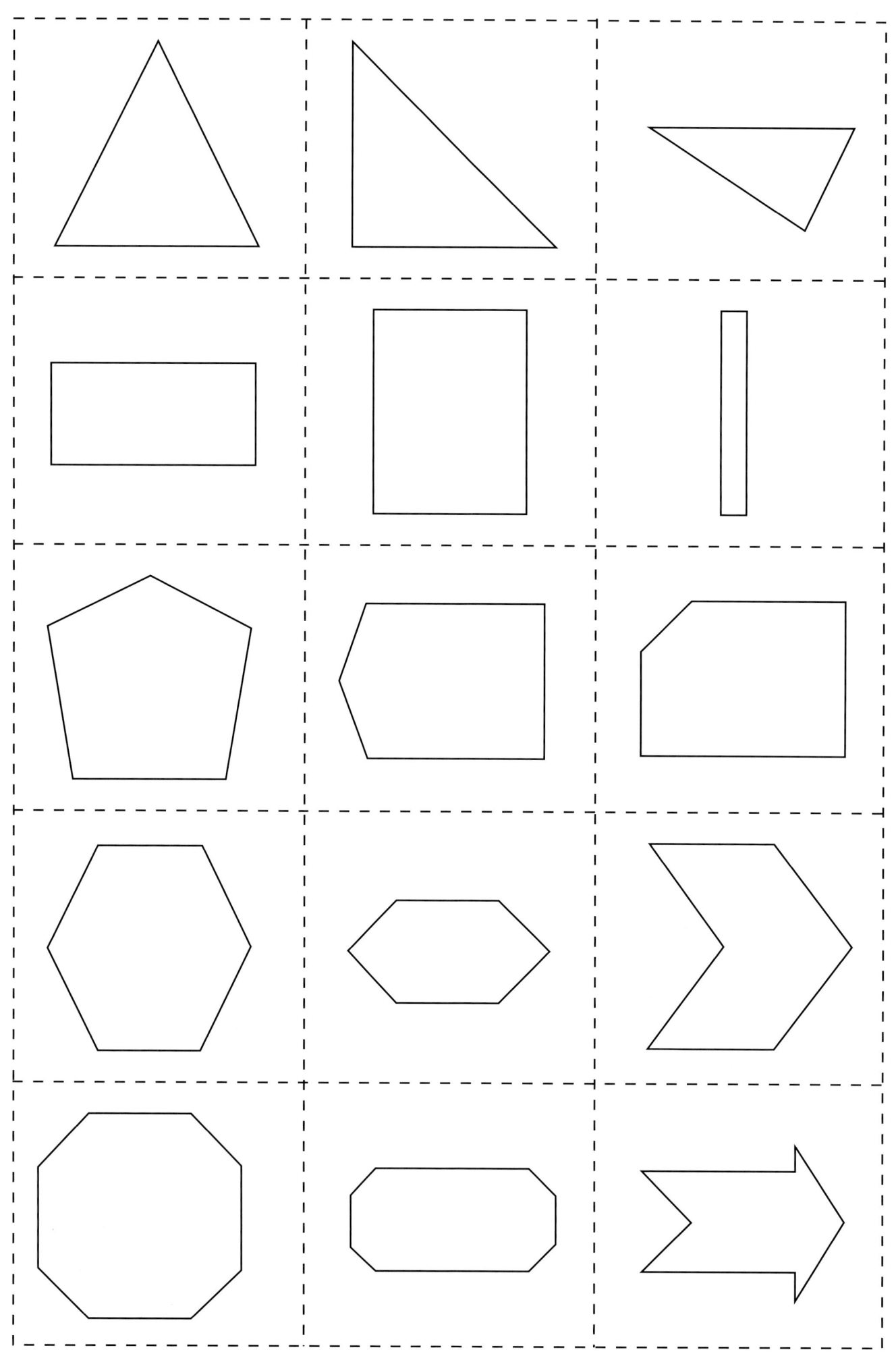

Shape cards

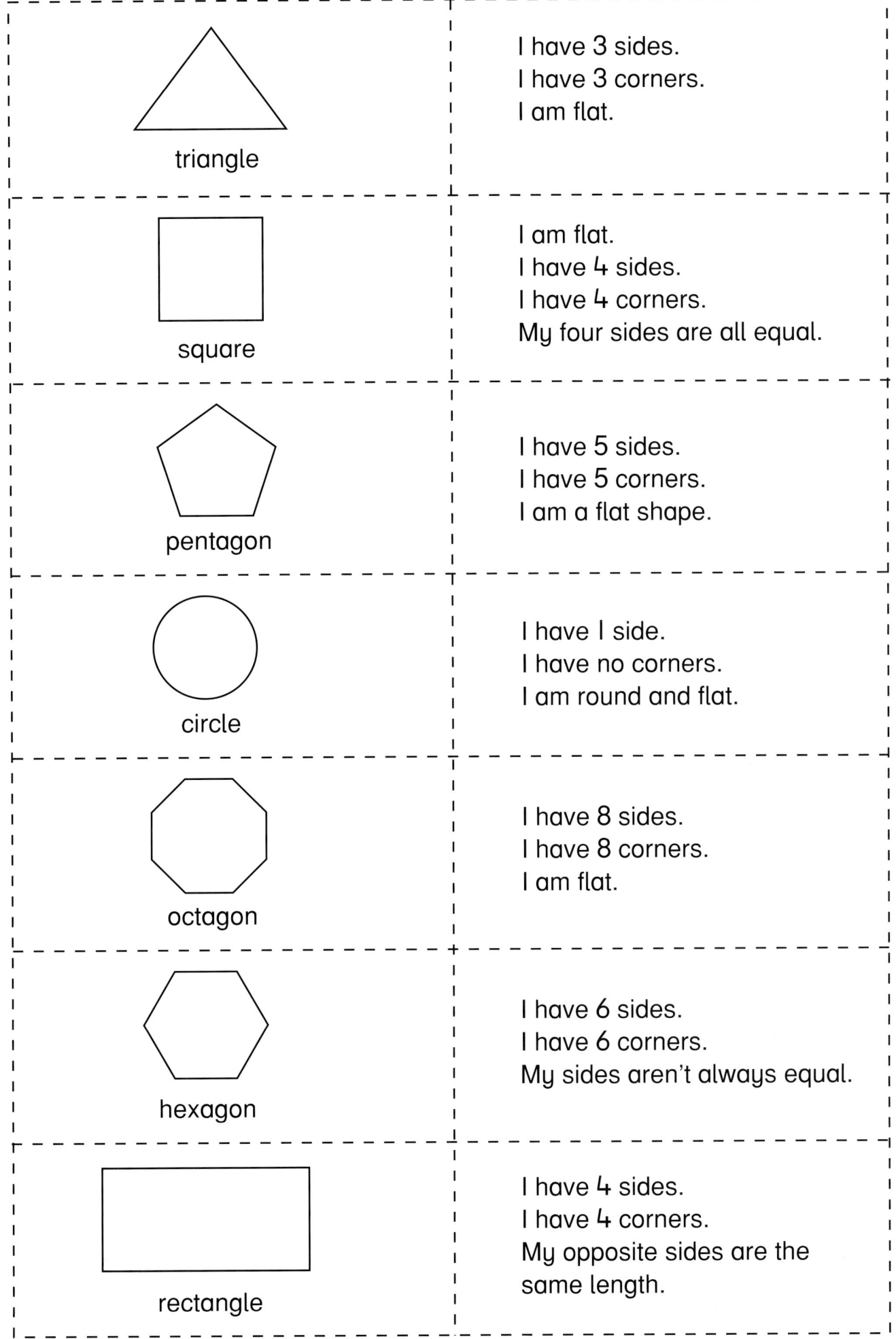

triangle	I have 3 sides. I have 3 corners. I am flat.
square	I am flat. I have 4 sides. I have 4 corners. My four sides are all equal.
pentagon	I have 5 sides. I have 5 corners. I am a flat shape.
circle	I have 1 side. I have no corners. I am round and flat.
octagon	I have 8 sides. I have 8 corners. I am flat.
hexagon	I have 6 sides. I have 6 corners. My sides aren't always equal.
rectangle	I have 4 sides. I have 4 corners. My opposite sides are the same length.

Flat shapes and their properties

Shape challenge

Shape house

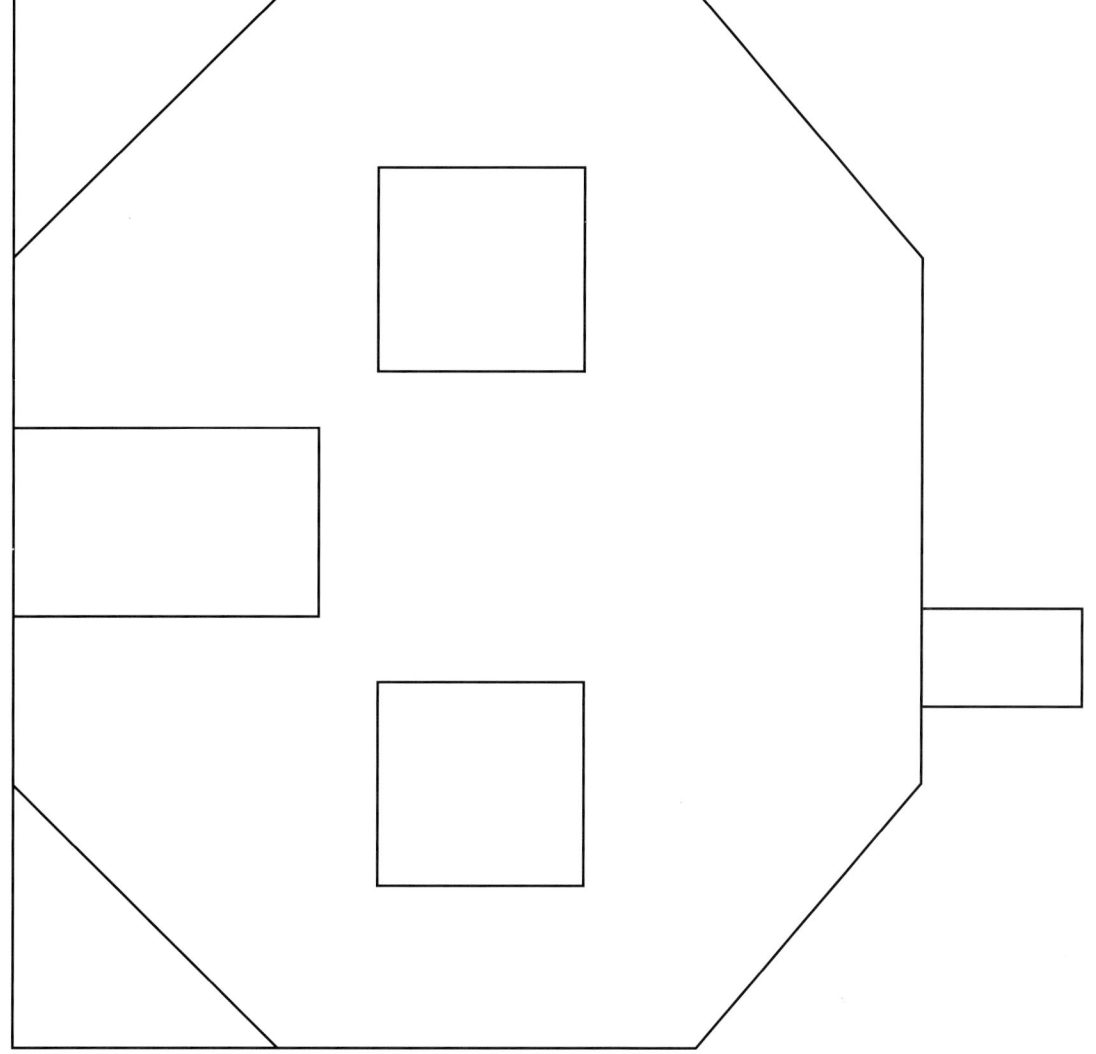

Can you draw these shapes?

Part a

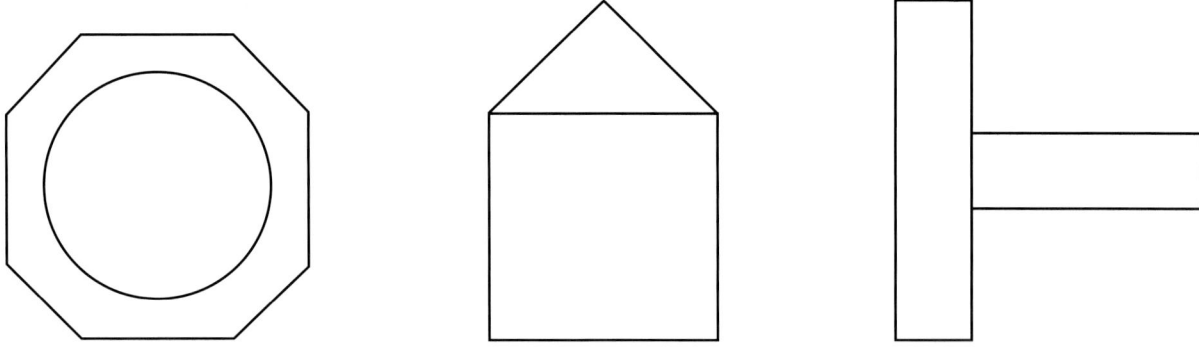

Describe what you see to your friend. Your friend will draw it.
Do your pictures look the same?

Part b

Describe what you see to your friend. Your friend will draw it.
Do your pictures look the same?

Cambridge Mathematics Direct 2 © Cambridge University Press 2002   SS1.3

Shape castle

- Take turns to take a card. Answer the question.
- Move to the next shape that matches your answer.
- The winner reaches the castle first.

You need

2–4 players
Cards from AS 60
1 counter for each player

Start

B1	B2	B3
I have 5 faces, 4 corners and a point. 4 of my faces are triangular. What am I?	I have a circular face, a curved face and a point. What am I?	I have 2 circular faces. I can roll. What am I?

B4	B5	B6
I have 6 square faces. All are flat. I have 12 straight edges. What am I?	My faces can be two shapes, square or rectangular. What am I?	I have no corners or edges. I can roll. What am I?

C1	C2	C3
How many curved faces do I have?	How many corners do I have?	How many faces do I have?

C4	C5	C6
How many edges do I have?	How many flat faces do I have?	How many circular faces do I have?

C7	C8	C9
How many triangular faces do I have?	How many rectangular faces do I have?	How many square faces do I have?

Shape castle game cards

I have 3 faces.
Two are circular.
The other is curved.
What am I?

I have 2 edges and
no corners.
I roll.
What am I?

I have I square face
and 4 triangular
faces.
All my faces are flat.
What am I?

I have 5 corners and
8 edges.
I do not roll.
What am I?

My faces can be
rectangular or square.
I have 12 edges.
What am I?

I have 6 faces and
8 corners.
My edges are not all
equal in length.
What am I?

I have 6 faces.
They are all equal
and square.
What am I?

I have 12 edges all
the same length.
I do not roll.
What am I?

I have I curved face.
I roll.
What am I?

I have no corners
and no edges.
What am I?

I have 2 faces.
One face is curved.
My other face is a
circle.
What am I?

I have a point.
I have I flat face.
I roll.
What am I?

What am I? game cards

Making cuboids

You need Multilink cubes.

(A) 1 Cover the shape on the right with cubes to make
a cuboid.
- Make your cuboid 1 cube tall.
- How many cubes?

2 Use the same cubes to make 3 more cuboids.
- Make them 1 cube tall.
- Draw around them.

62

Making solid shapes

Name:

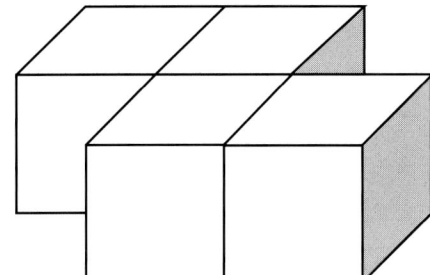
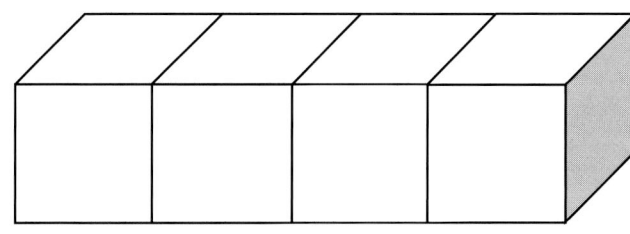

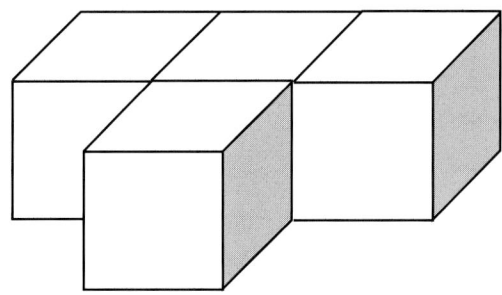

Make these shapes.

Use 4 cubes to make 2 more shapes. Draw them.

Find the shapes

Name:

You need coloured pencils.

(A) 1 Colour

cubes	red	cuboids	green
cylinders	blue	cones	orange
spheres	purple	pyramids	yellow

2 How many

a cubes? ☐

b cuboids? ☐

c cylinders? ☐

d cones? ☐

e spheres? ☐

f pyramids? ☐

Shape patterns

Ⓒ **1** Make arrays of cylinders like these.

a Start with 2. **b** Start with 3. **c** Start with 4.

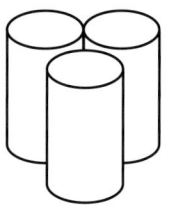 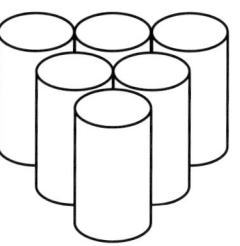 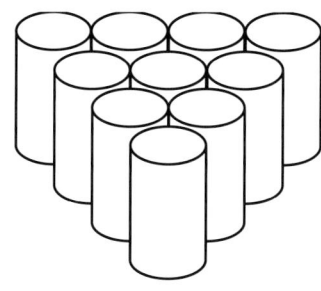

How many cylinders? How many cylinders? How many cylinders?

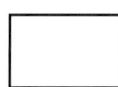 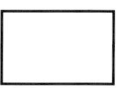 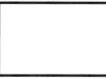

d Start with 5. How many cylinders?

e Start with 6. How many cylinders?

f Can you carry on the number pattern? It starts like this.

3, 6, 10, ☐ , ☐ , ☐ , ☐

2 **You need** building blocks.
- Build a wall.
- Make it bigger. Think of a rule.
- Investigate number patterns.

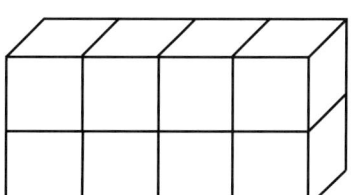

Notes/date:

Symmetry

2 Finish the pattern here.
Make it symmetrical.

Use plastic shapes.
1 Make a pattern here.

Notes/date:

Symmetrical shapes?

Name:

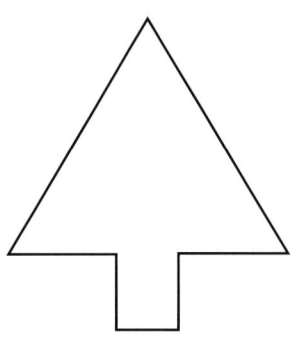

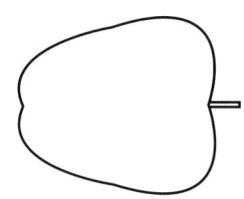

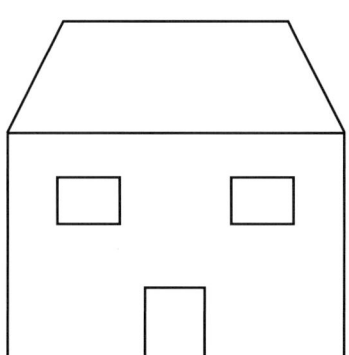

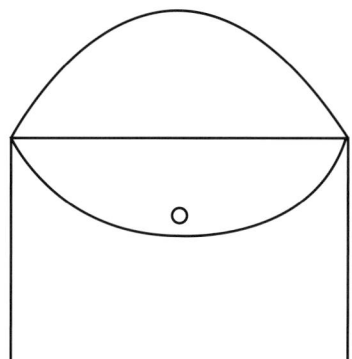

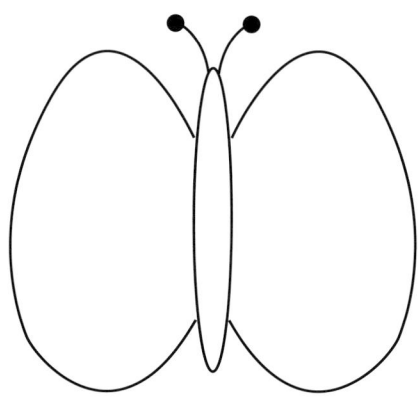

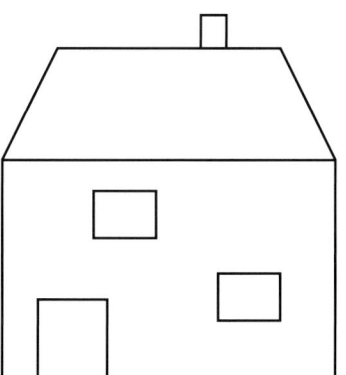

Cambridge Mathematics Direct 2 © Cambridge University Press 2002   SS 3.2

Make symmetrical

B 1

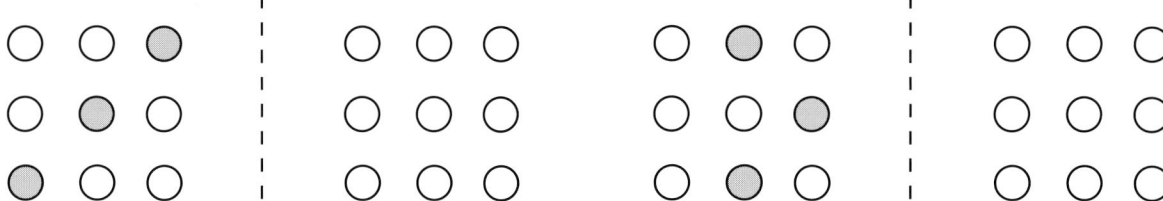

Check with a mirror.

2 Use a mirror to help you complete each face to make it symmetrical.

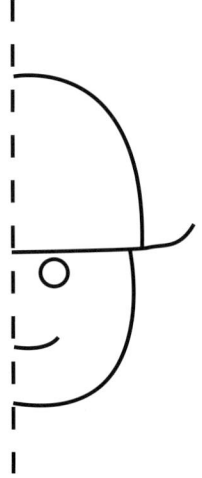

The playground

Name:

(A)

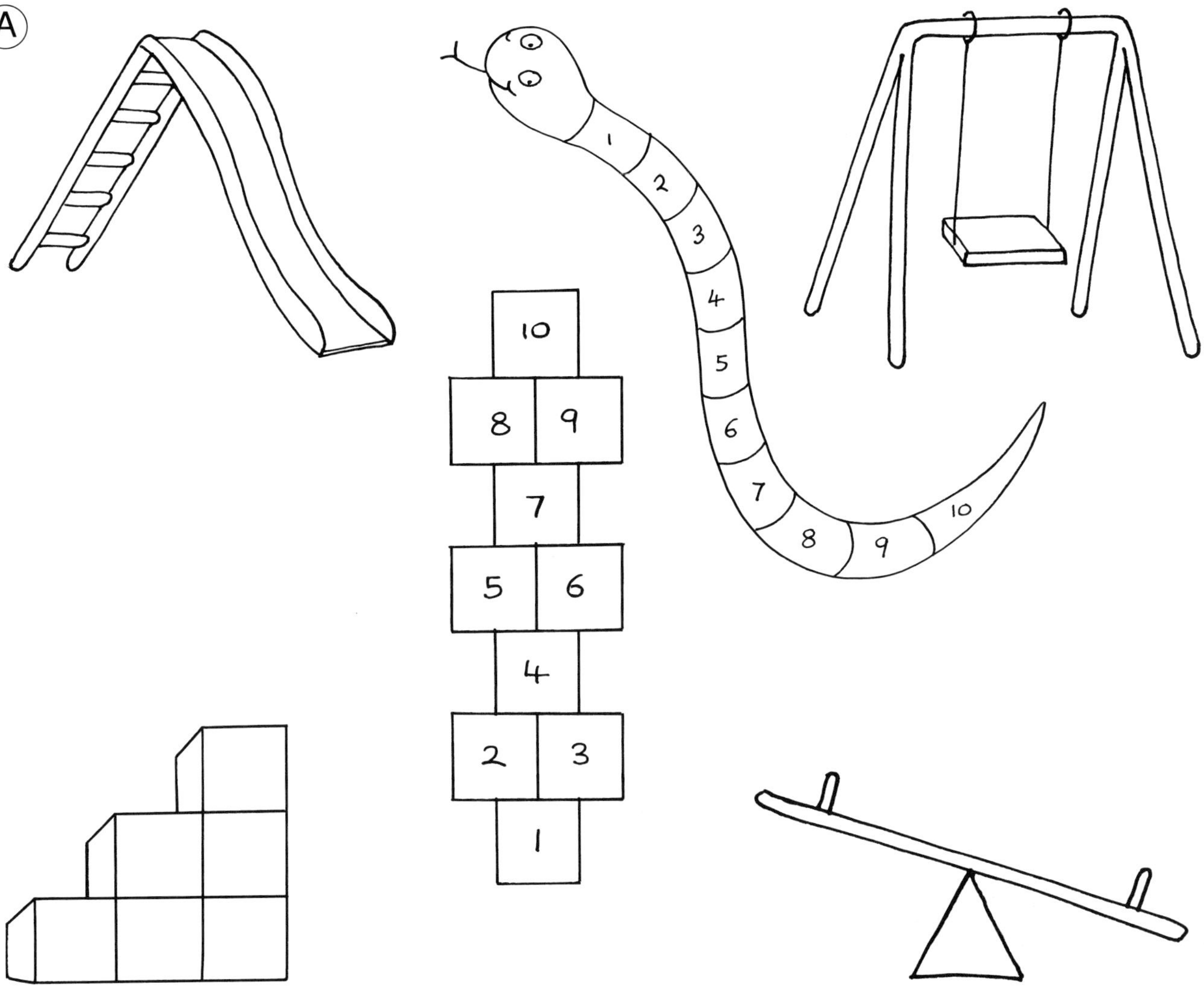

Put Jack under the  slide.

Put Tom on top of the building blocks.

Put Lizzie beside the swing.

Put Leanne in front of the 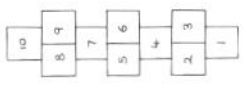 hopscotch.

Put Omar between the seesaw and the swing.

Jack

Tom

Lizzie

Leanne

Omar

Notes/date:

Cambridge Mathematics Direct 2 © Cambridge University Press 2002   SS4.1

Following routes

4				12		
	9				100	
		2				
18						
				15		50
	16					
	0			20		
						10

Cambridge Mathematics Direct 2 © Cambridge University Press 2002   SS 4.2

The maze

B

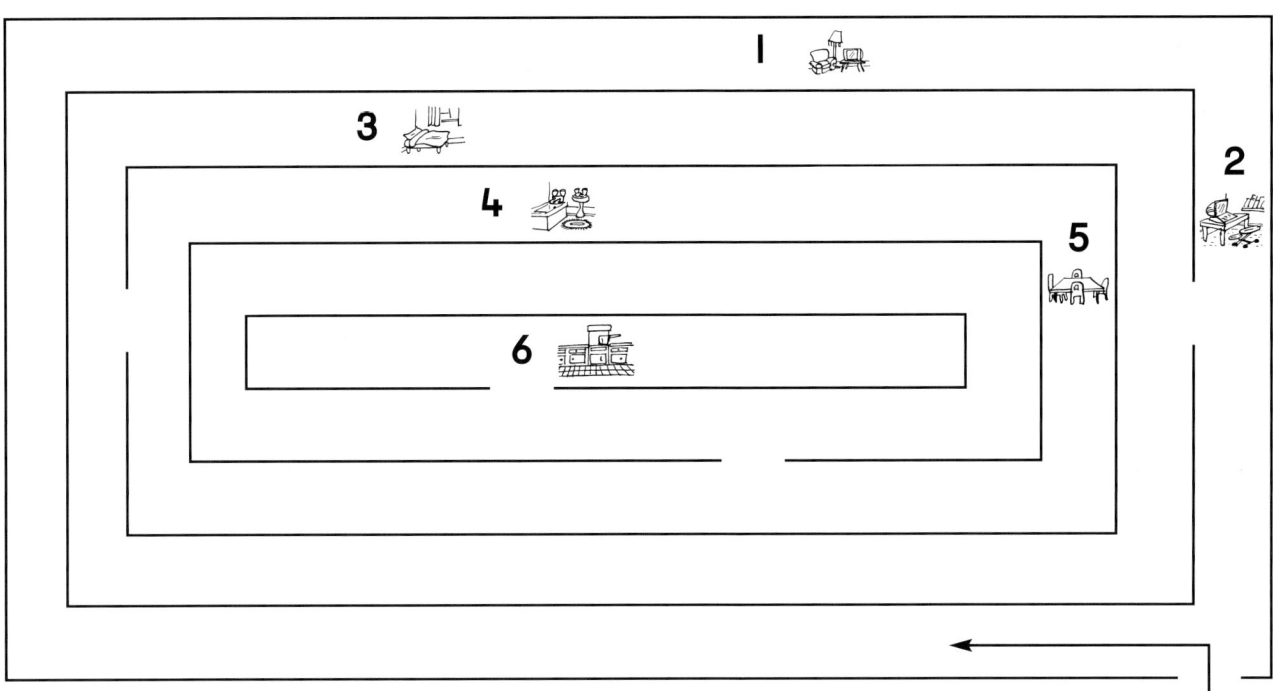

1 Draw a route that passes through each room, in order from here: start

   Finish at ![icon] .

2 How many times do you turn, clockwise or anti-clockwise, between the rooms?
   Fill in the table.

Room	Number of turns	Number clockwise	Number anti-clockwise
![icon]	3	2	1
![icon]			
![icon]			
![icon]			
![icon]			
![icon]			

3 How many turns do you make altogether to reach ?

Half and quarter turns

Name:

(A) Use geo-strips to help you.
Draw a quarter turn and a half turn from each starting position.

Start	Quarter turn	Half turn
	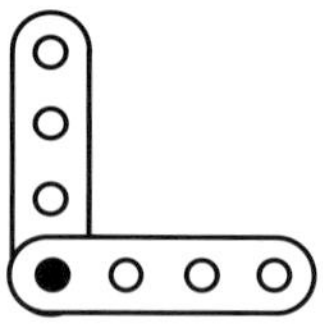	

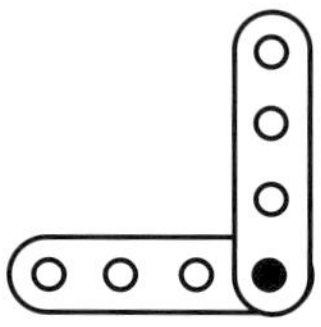		
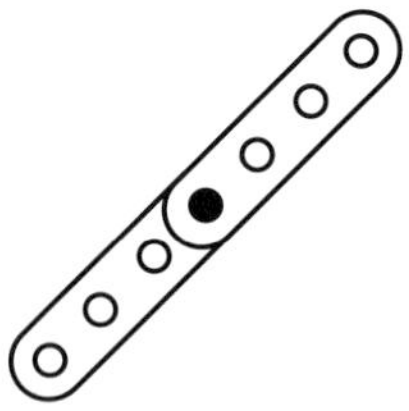		
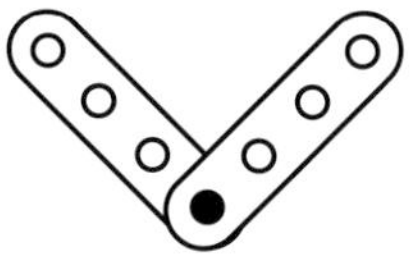		

Notes/date:

Moving shapes

(B) Draw the next clockwise quarter turns.
Use cubes or shapes to help you.

1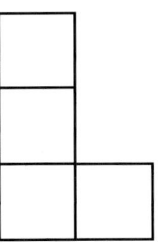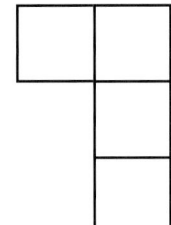

2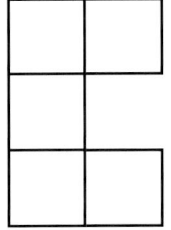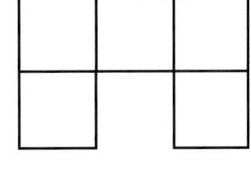

3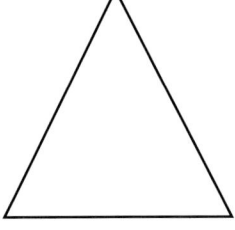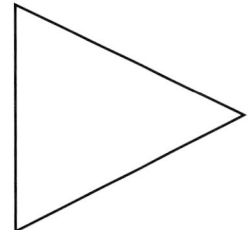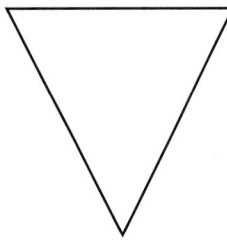

4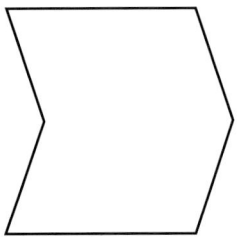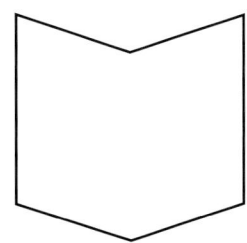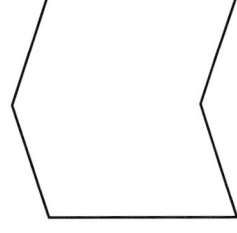

5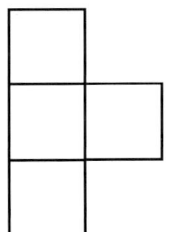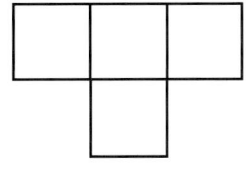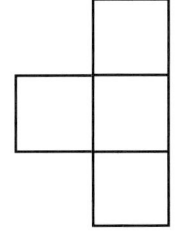

6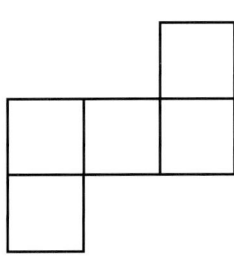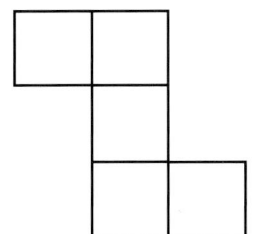

Titles for lists

(A)   multiples of 10 between 0 and 50

(A)   ten numbers greater than 70

(A)   8 even numbers less than 25

(B)   multiples of ten between 40 and 110

(B)   odd numbers between 8 and 20

(B)   2-digit numbers ending in 5

(★)   odd numbers from 1 to 10

(★)   numbers from 5 to 18

(★)   ten numbers greater than 6

Lists of numbers

1	2	3	4	5	6	7	8	9	10
11	12	13	14	15	16	17	18	19	20
21	22	23	24	25	26	27	28	29	30
31	32	33	34	35	36	37	38	39	40
41	42	43	44	45	46	47	48	49	50
51	52	53	54	55	56	57	58	59	60
61	62	63	64	65	66	67	68	69	70
71	72	73	74	75	76	77	78	79	80
81	82	83	84	85	86	87	88	89	90
91	92	93	94	95	96	97	98	99	100

Make a list of:

Stick your title here.

Cambridge Mathematics Direct 2 © Cambridge University Press 2002   HD1.1

How much paper?

Name:

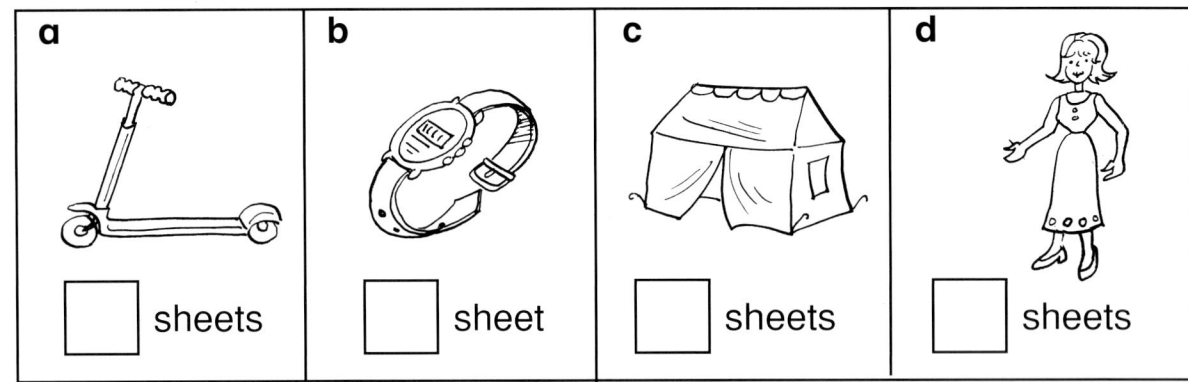

(A)

You need AS 77.

1

a	**b**	**c**	**d**
☐ sheets	☐ sheet	☐ sheets	☐ sheets

2 **a** Choose a toy that takes 1 sheet. _____

   b Choose a toy that takes 3 sheets. _____

3  and  ☐ sheets altogether

(B) 4 Write your own question.

_____

_____

5 **a**  and ☐ sheets altogether

   b 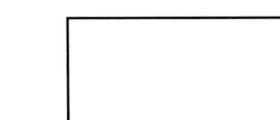 ☐ sheets

   c 5 sheets: choose 2 presents

      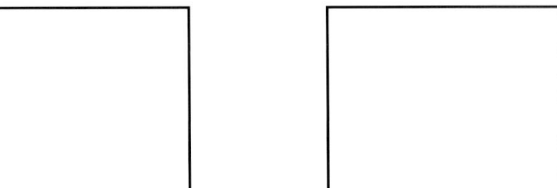

6 All the presents! How many sheets? ☐

Wrapping presents

How many sheets of paper to wrap the presents?

Paper	Presents		
1 sheet	books	watch	walkie-talkie set
2 sheets	doll	football strip	play station
3 sheets	snooker table	scooter	model
4 sheets	bike	tent	drums

Name(s):

Pictogram

78

Party pictogram

Cambridge Mathematics Direct 2 © Cambridge University Press 2002   HD 2.1

Happy faces

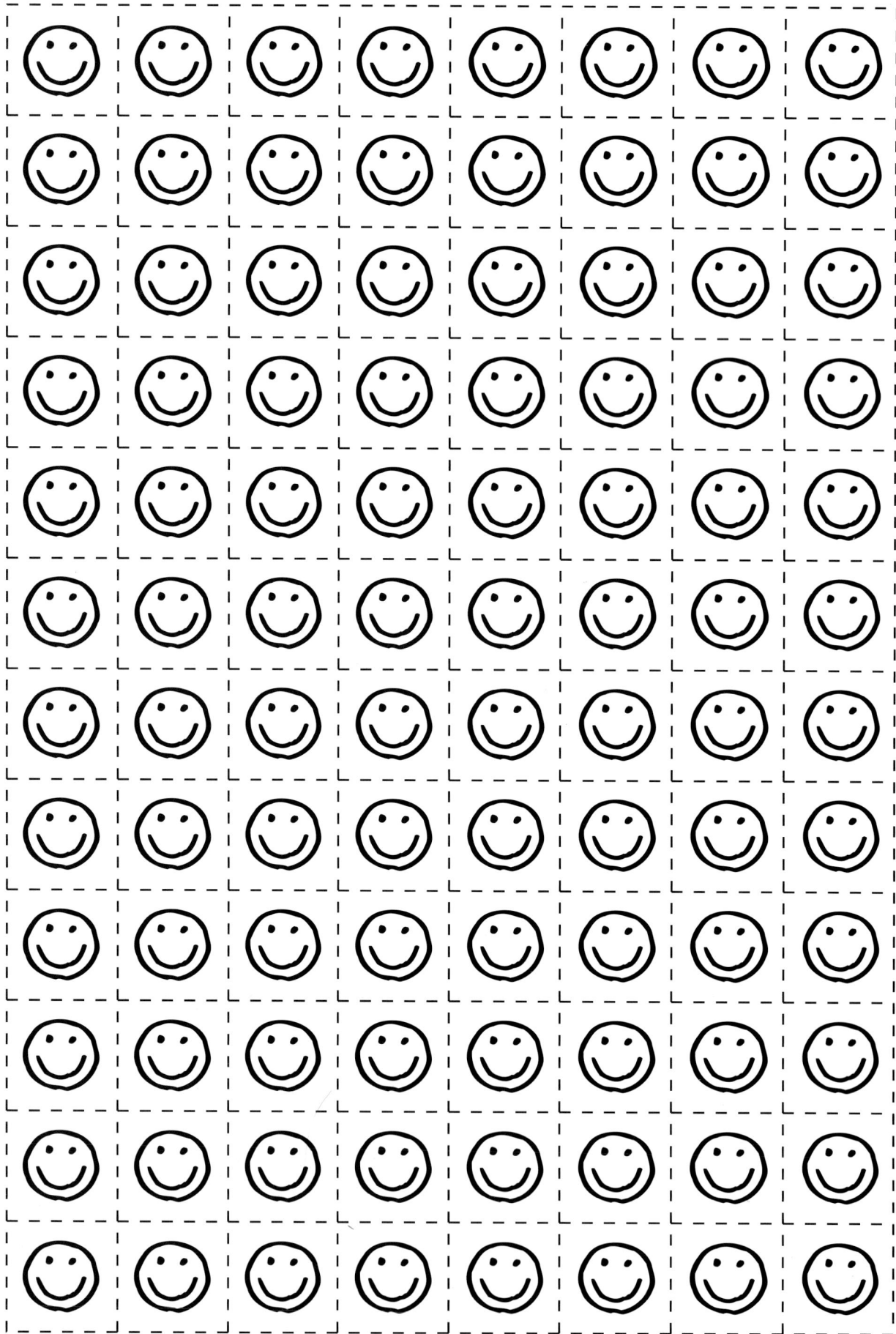

Block graph grid

Name:

Title of graph: _____

Notes/date:

Cambridge Mathematics Direct 2 © Cambridge University Press 2002   HD 3.1, 3.3

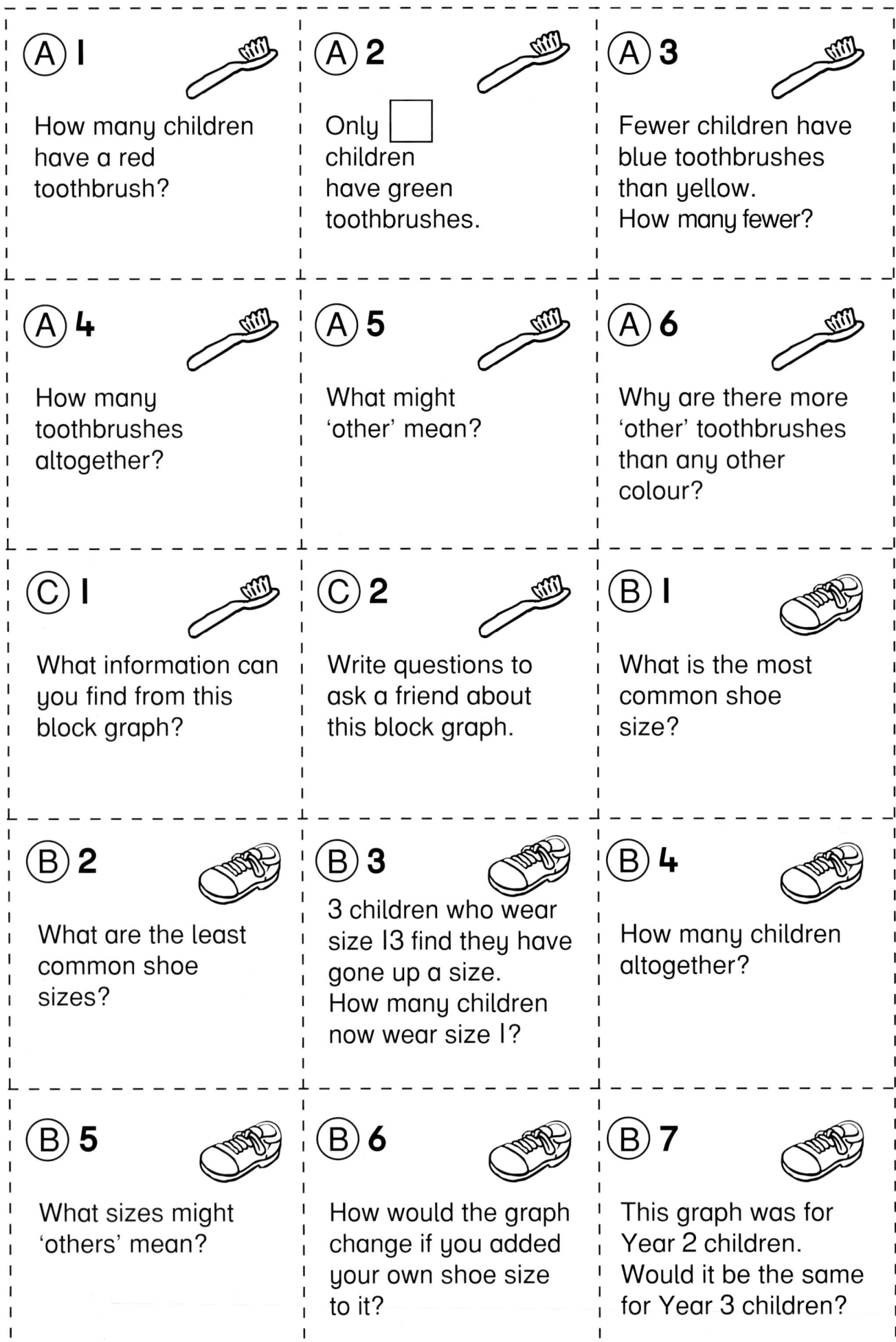

A 1
How many children have a red toothbrush?

A 2
Only ☐ children have green toothbrushes.

A 3
Fewer children have blue toothbrushes than yellow. How many fewer?

A 4
How many toothbrushes altogether?

A 5
What might 'other' mean?

A 6
Why are there more 'other' toothbrushes than any other colour?

C 1
What information can you find from this block graph?

C 2
Write questions to ask a friend about this block graph.

B 1
What is the most common shoe size?

B 2
What are the least common shoe sizes?

B 3
3 children who wear size 13 find they have gone up a size. How many children now wear size 1?

B 4
How many children altogether?

B 5
What sizes might 'others' mean?

B 6
How would the graph change if you added your own shoe size to it?

B 7
This graph was for Year 2 children. Would it be the same for Year 3 children?

Reading block graphs

Cambridge Mathematics Direct 2 © Cambridge University Press 2002   HD 3.2

0	1	2	3	4	5	6	7	8
9	10	11	12	13	14	15	16	17
18	19	20	21	22	23	24	25	26
27	28	29	30	31	32	33	34	35
36	37	38	39	40	41	42	43	44
45	46	47	48	49	50	51	52	53

0–53 number line   Cambridge Mathematics Direct 2 © Cambridge University Press 2002

G

47	48	49	50	51	52	53	54	55
56	57	58	59	60	61	62	63	64
65	66	67	68	69	70	71	72	73
74	75	76	77	78	79	80	81	82
83	84	85	86	87	88	89	90	91
92	93	94	95	96	97	98	99	100

47–100 numberline

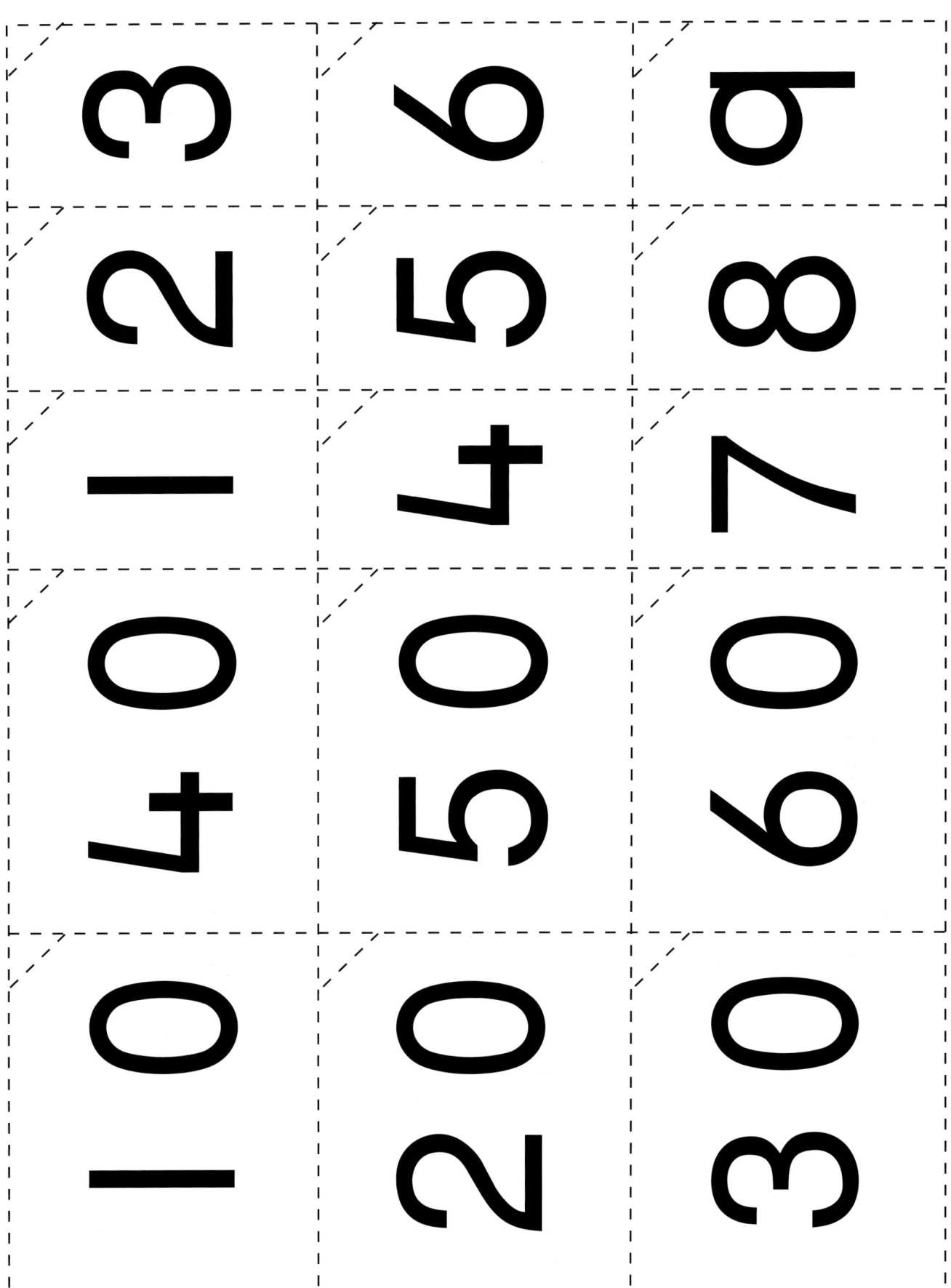

Place value cards 1

Cambridge Mathematics Direct 2 © Cambridge University Press 2002

G

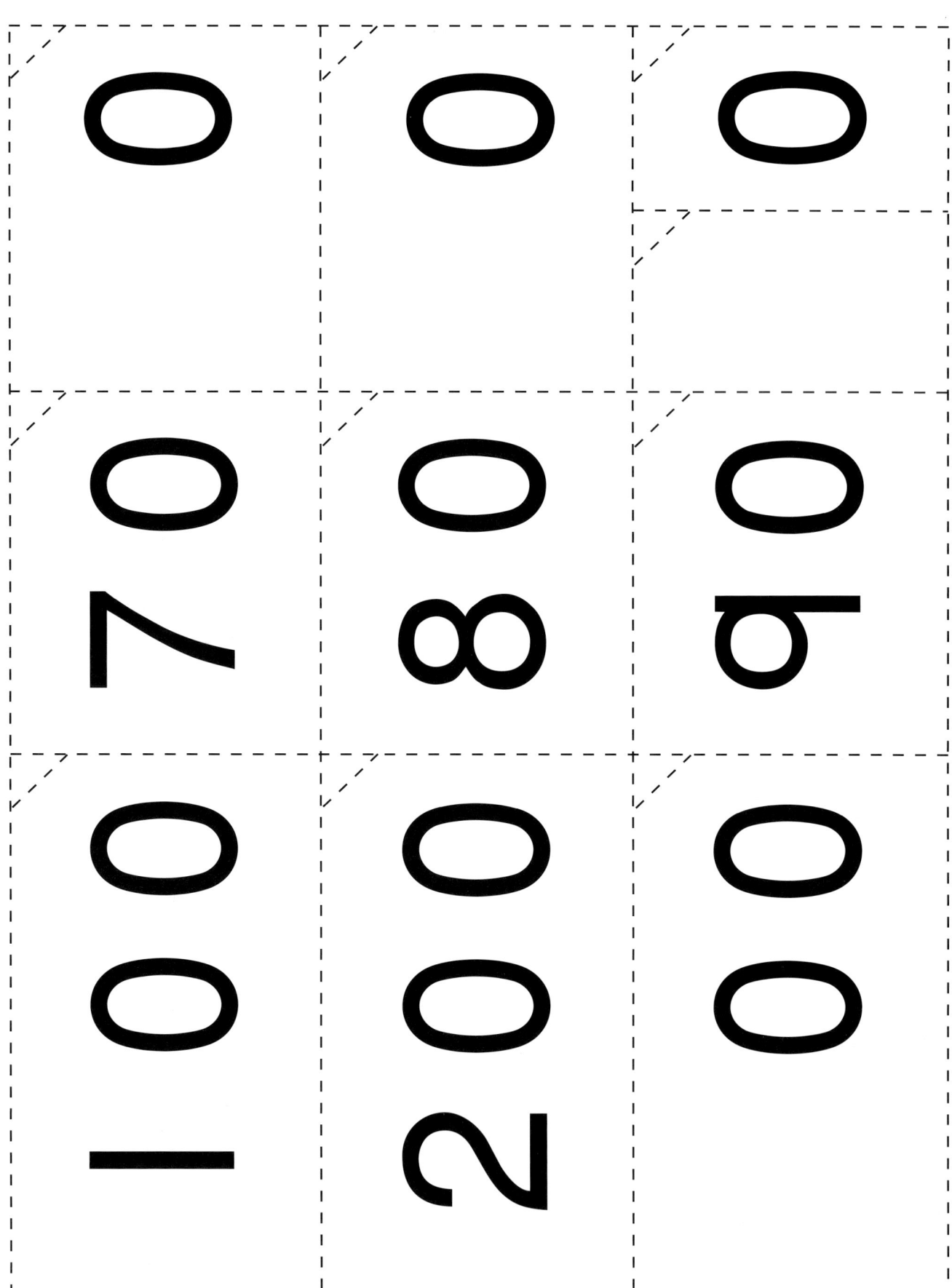

Place value cards 2

Cambridge Mathematics Direct 2 © Cambridge University Press 2002

Blank weighing scales

Cambridge Mathematics Direct 2 © Cambridge University Press 2002

G

Measuring jug

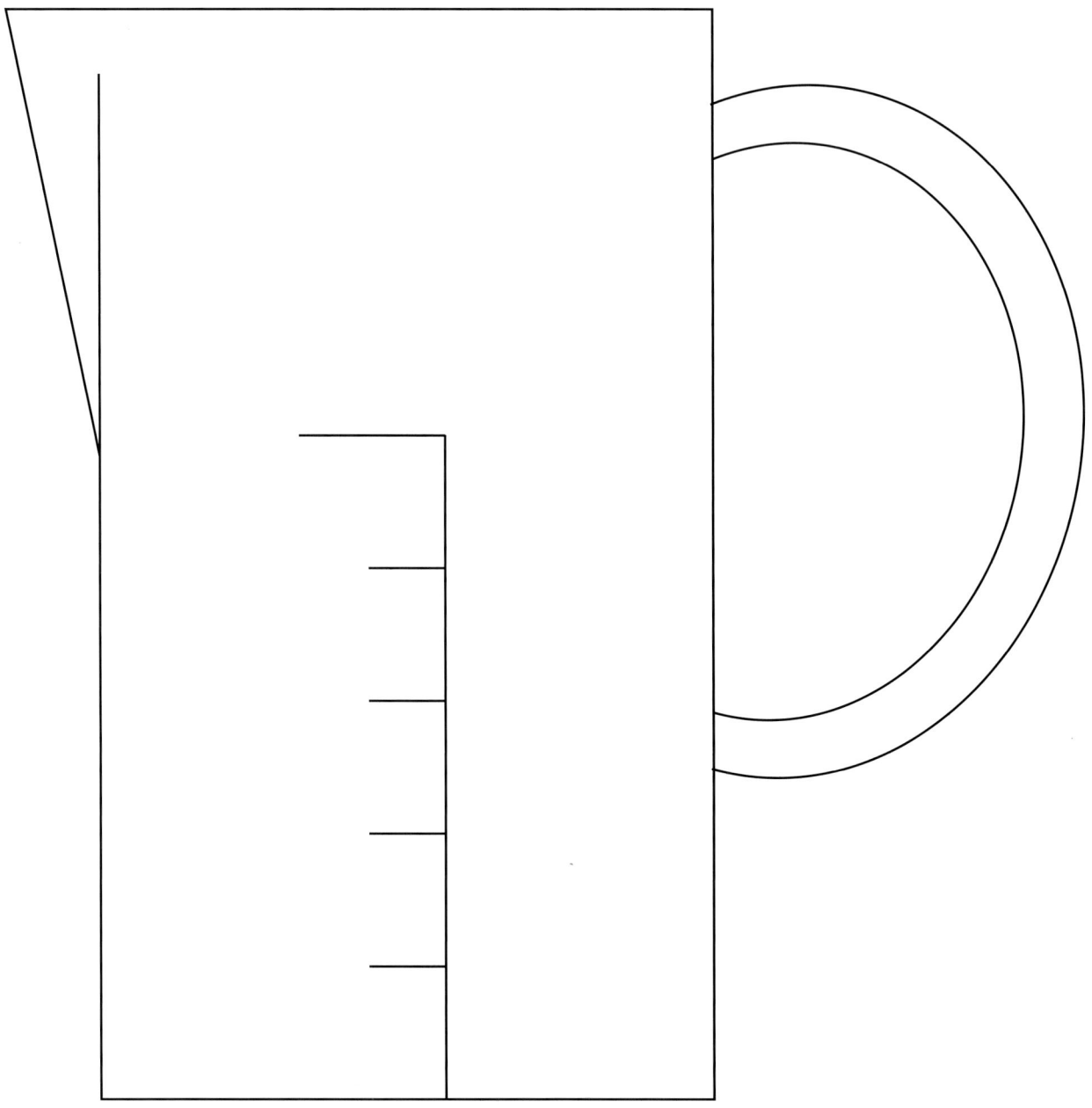